Classical
Sociological
Theory

Classical Sociological Theory

A Positivist's Perspective

Jonathan H. Turner
University of California—Riverside

Nelson-Hall Publishers
Chicago

Library of Congress Cataloging-in-Publication Data

Turner, Jonathan H.
    Classical sociological theory : a positivist's perspective /
Jonathan Turner.
      p.   cm.
    Includes bibliographical references (p.   ) and index.
    ISBN 0-8304-1349-9
    1. Sociology—Philosophy. 2. Sociology—History. 3. Positivism.
I. Title.
HM24.T837   1993
301'.01—dc20   92-31256
                                                        CIP

Manufactured in the United States of America

10 9 8 7 6 5 4 3 2 1

 ™ The paper used in this book meets the
minimum requirements of American
National Standard for Information
Sciences—Permanence of Paper for
Printed Library Materials, ANSI
Z39.48-1984.

To My Good Friend,

Richard Rosenberg

Contents

Preface   ix

Acknowledgments   xi

1. Comte Would Turn Over in His Grave   1

2. Toward a Social Physics: Reducing Sociology's
   Theoretical Inhibitions   7

3. The Forgotten Theoretical Giant: Herbert Spencer   17

4. Spencer's Human Relations Area Files   35

5. Émile Durkheim's Theory of Integration in
   Differentiated Social Systems   47

6. Durkheim's and Spencer's Principles of Social
   Organization   57

7. Émile Durkheim's Final Theory of Social
   Organization   67

8. Marx and Simmel: Reassessing the Foundations of
   Conflict Theory   87

9. Where Marx Went Wrong   101

10. Max Weber's Theory of Integration and Conflict   109

11. Marx and Weber Meet the Modern-day Adam Smith   119

12. Spencer, Marx, Weber, Simmel, and Pareto on Power
    and Conflict   135

13. George Herbert Mead's Behavioral Theory of Social
    Structure   149

14. Mead as a Social Physicist   159

15. What Went Wrong? Detours from the Early Masters   173

References   191

Index   201

Preface

I used to be highly suspicious of authors' collecting their own essays. Were they creating a lazy person's book, or trying to scrape together some needed cash? But over the years, I have found some collections very useful, allowing the reader to examine another scholar's ideas in a more readily available format. The key to a good book of essays is this: Do the essays "hang together" and develop a line of argument or a set of themes?

This question has guided the collection of essays in this book. All the essays are devoted to one theme: Sociology can be a true science, and it can develop abstract laws explaining the operative dynamics of the social universe. I have executed this theme by pulling together articles on sociology's early masters in which I tried to formalize their theories and to demonstrate the power of those theories to explain social phenomena. I also wrote several new essays in order to fill in gaps and give the book coherence and continuity.

There is some overlap in the essays, but this is all to the good: It drives home my central theme. There is also a polemical tone to the essays that may irritate some who do not agree with me, but that for most should make for more interesting reading.

This book is dedicated to one goal: to codify the wisdom of the masters so that we can move on and make books on classical theory unnecessary. A true science incorporates the ideas of its early founders in introductory texts and moves on, giving over the analysis of its founders to history and philosophy. Sociology, on the other hand, has stagnated and become entangled with its early masters, worshiping them and canonizing the sacred texts of St. Marx, St. Weber, St. Durkheim, and St. Mead. We should have pulled the key theories from these scholars long ago, and then moved on. But instead, we read and reread the sacred texts and engage in incessant and constant discourse about them. A science whose theorists worship text, per se, does not accumulate knowledge; it proliferates verbiage that does not tell us how the social universe operates.

Thus, in these essays, I have tried to extract those portions of classical theory that tell us how the social world works. The rest of the early theorists' work, plus the usual biography, philosophy, and history of ideas, I have discarded. It is the abstract laws of human interaction and organization revealed in the classical works that should guide our reading of them, not those issues best left to historians and philosophers. In this book, therefore, I present my best effort to extract these laws; I invite others to do the same and to compare their notes with mine. In this way, we can use discourse over the early classics to build a science of society.

Jonathan Turner
Idyllwild, California

Acknowledgments

I want to thank all my students who, for the last twenty-five years, have endured my polemical lecturing style. They have giggled and rolled their eyes at many pronouncements, but at the same time they seem to have enjoyed a more forceful and unrestrained mode of presentation. In these essays, I have been true to this lecturing style (or as many have said, "preaching style"), abandoning many of the conventions of dull academic prose. I trust others will enjoy this more polemic approach.

I am also grateful to the following publishers who have given me permission to reprint all or portions of various articles:

- The JAI Press, 55 Old Post Road, PO Box 1678, Greenwich, CT 06830 which holds the copyrights for the materials presented in chapters 4, 5, 6, 12, and 14.
- *The Humboldt Journal of Social Relations*, Department of Sociology, Humboldt State University, Arcata, CA 95521 which holds the copyright for the material in chapter 2.
- *Social Forces*, Department of Sociology, University of North Carolina, Chapel Hill, NC 27599 which holds the copyrights for the material in chapters 7 and 8.
- Blackwell, 108 Cowley Road, Oxford OX4 1JF, England which holds the copyright for the materials in chapter 13.

1 Comte Would Turn Over in His Grave

Auguste Comte is a quietly ridiculed figure in sociology. True, we still honor him for giving us our name—the Latin-Greek hybrid *sociology*—but we never forget that he was also the man who, near the end of his career, proclaimed himself the "Founder of Universal Religion" and "The Great Priest of Humanity" (Comte 1851–1854). How are we to take seriously such a man? And should we care if, as the title of this opening chapter declares, he turns over in his grave because of our wanderings from his preachings? Indeed, Comte seems a bit of an embarrassment to sociology, and, in the view of many, it would be just as well to forget him and his work. Yet the early Comte—the one who wrote *Cours de philosophie positive*—is worth remembering, especially in the opening pages of this four-volume work in which he argues for a particular view of science.

AUGUSTE COMTE AND POSITIVISM

In *Cours de philosophie positive*, Comte lays out his views on positivism, and these ideas are worth repeating. The basic message is that "the first characteristic of Positive Philosophy is that it regards all phenomena as subject to invariable natural *Laws*. . . . Our real business is to analyze accurately the circumstances of phenomena, and to connect them by the natural relations of succession and resemblance" (Comte 1830, 5–6). Comte thought that Newton's famous equation on gravitation "was the best illustration" of what he had in mind for sociology, and perhaps for this reason, he preferred the label *social physics* to the Latin-Greek hybrid that he was forced to accept.[1] Comte was probably shooting too high in

Small portions of this chapter appear in Jonathan H. Turner, "Positivism," in the *Encyclopedia of Sociology*, ed. E. Borgatta (New York: Macmillan, 1991).

believing that sociology could fully emulate physics (biology would be a better model), but he was certainly right on the mark with his insistence that sociology can develop abstract laws and principles that allow us to understand the operative dynamics of the social universe. Such theoretical principles must be testable, for "every theory must be based upon observed fact." In arguing this way, Comte was probably aware of what theory becomes—metatheory and philosophizing—when it is detached from considerations of the empirical world. Much contemporary social theory would, in Comte's view, represent a regression to the "metaphysical" stage of intellectual development, where ideas and systems of thought are constructed with few efforts at an empirical reality check.

Comte was also to anticipate another great flaw of contemporary sociology: atheoretical empiricism or, more pejoratively, "mindless empiricism." For Comte (1830, 4), "if it is true that every theory must be based upon observed facts, it is equally true that facts cannot be observed without the guidance of some theory." Moreover, along with theory that is not oriented to explaining empirical events, the "next great hindrance" to sociology is "empiricism which is introduced into it by those who, in the name of impartiality, would interdict the use of any theory whatever" (Comte 1830, 242). For "no dogma could be more thoroughly irreconcilable with the spirit of positive philosophy"; indeed, "no real observation of any phenomena is possible, except in so far as it is first directed, and finally interpreted, by some theory" (Comte 1830, 243).

How, then, did positivism become identified in the contemporary mind with "raw empiricism" and atheoretical work? There is, I think, a nice dissertation for someone on this remarkable transposition of Comte's intent, but let me offer a cursory assessment of what I think happened.

WHAT HAPPENED TO POSITIVISM?

Much as Marx "stood Hegel on his head," so sociology has turned Comte's ideas upside down. How did this occur? The answer to this question cannot be found in nineteenth-century sociology, for the most positivistic sociologists of this period—Herbert Spencer (1874–1896) and Émile Durkheim (1893, 1895)—could hardly be accused of "raw" and "mindless" empiricism. Moreover, early American sociologists—Albion Small, Frank Lester Ward, Robert Park, William Graham Sumner, and even the father of statistical methods and empiricism in American sociology, Franklin Giddings—all advocated Comtean and Spencerian positivism before World War I. Thus, the answer to this question is to be found in the natural sciences, particularly in a group of scientist-philosophers who are sometimes grouped under the rubric "the Vienna Circle," de-

spite the fact that several intellectual generations of very different thinkers were part of this "circle."

Anticipating the debates of the Vienna Circle, Ernst Mach (1893)
argued that the best theory employs a minimum of variables and does not
speculate on unobservable processes and forces. Mach emphasized reliance on immediate sense data, rejecting all speculation about causes and
mechanisms to explain observed relations among variables. Indeed, he
rejected all conceptions of the universe as being regulated by "natural
laws" and insisted that theory must represent mathematical descriptions
of relations among observable variables. Mach's ideas framed many of the
issues later debated within the Vienna Circle, but did not dictate their
resolution. Some thinkers in the circle were primarily concerned with
logic and systems of formal thought to the exclusion of observation (or, at
least, to its subordination). A split thus developed in the circle over the
relative emphasis on empirical observation and systems of logic: A radical
faction emphasized that truth can be "measured solely by logical coherence of statements" (which had been reduced to mathematics), whereas a
more moderate group insisted that there is a "material truth of observation" supplementing "formal truths" (Johnston 1983, 189). Karl Popper,
who was at the periphery of the late Vienna Circle (of the 1930s), is perhaps the best known mediator of this split, for he clearly tried to keep the
two points of emphasis together. But even here the reconciliation is
somewhat negative (Popper 1959, 1969): A formal theory can never be
proven, only disproven; and so, data are used to mount assaults on abstract theories from which empirical hypotheses and predictions are formally "deduced."

Why did the philospher-scientists in the Vienna Circle have any impact on sociology, especially American sociology? In Europe, of course,
sociology had always had a foot (if not a leg and part of the torso) firmly
anchored in philosophy, but in American sociology during the 1920s and
1930s, the rise of quantitative sociology was accelerating as the students
of Franklin Giddings assumed key positions in academia and as Comtean
and Spencerian sociology became a distant memory. (It should be noted,
however, that Marx, Weber, and Durkheim had yet to have much impact
on American sociology in the late 1920s or early 1930s). But American
sociology was concerned with its status as science and, hence, was recep-
tive to philosophical arguments that could legitimate its scientific aspirations (Turner and Turner 1990). Mach was appealing because his advocacy legitimated statistical analysis of empirical regularities as variables;
and Popper was to win converts with his uneasy reconciliation of observation and abstract theory. Both legitimated variable analyses; and for
American sociologists in the 1930s and later in the 1940s, 1950s, and early
1960s, this meant sampling, scaling, and statistically aggregating and ana-

lyzing empirical "observations." Members of the Vienna Circle had even developed an appealing term, "logical positivism," to describe this relation between theory (abstract statements organized by a formal calculus) and research (quantitative data for testing hypotheses logically deduced from abstract statements). The wartime migration of key figures in the late Vienna Circle to the United States no doubt increased their impact on the social sciences in the United States (despite the fact that the "logical" part of this new label for "positivism" was redundant in Comte's original formulation). But logical positivism legitimated American empiricism in this sense: The quantitative data could be used to "test" theories, and so it was important to improve upon data gathering and analyzing methodologies in order to realize this lofty goal. Along the way, the connection between theory and research was mysteriously lost; and positivism became increasingly associated with empiricism and quantification per se.

There was a brief and highly visible effort, reaching a peak in the late 1960s and early 1970s, to revive the "logical" side of positivism by explaining to sociologists the process of "theory construction." Indeed, numerous texts on theory construction were produced (e.g., Zetterberg 1965; Dubin 1969; Blalock 1969; Reynolds 1971; Gibbs 1972; Hage 1972), but the somewhat mechanical, cookbook quality of these books won few converts, and so the empiricist connotations of the term *positivism* were never successfully reconnected to abstract theory. Even the rather odd academic alliance of functional theory with quantitative sociology—for example, Merton-Lazarsfeld at Columbia and Parsons-Stouffer at Harvard—was unsuccessful in merging theory and research, once again leaving the label *positivism* to denote quantitative research divorced from theory.

In the rejection of the sociological effort to adopt logical positivism, Comte's vision was further obscured, or, in any event, was taken in an unhealthy direction. While sociology can articulate abstract laws and principles, the process of connecting them to empirical reality rarely occurs in the way that the various theory construction texts argued it occurred. Much of the logic and deduction advocated in these books is folk logic and folk deduction, in this sense: The law is used to guide a search for conditions approximating the variables in the law; measures of variables and relations are found that seem "reasonable"; and then, the theory is assessed. Only infrequently is a formal calculus—mathematics or symbolic logic—used to move from law to empirical hypothesis; and so on this score, the sociological version of logical positivism seems out of touch with actual practice, not just in sociology but also in most other sciences (with the exception of portions of physics). Thus, it was easy to reject, or ignore, the unrealistically rigid guidelines of the theory construction movement—a movement, I suspect, that Comte would have seen as more appropriate to a "metaphysical stage" of sociological development.

Comte's positivism became so distorted that it was associated with *both* raw empiricism and overconcern with logical deduction. The *raw empiricism* label has been the more enduring, but a number of theorists[2] prefer not to use the label *positivistic* to describe their work, because it smacks of the machinations of the Vienna Circle.

And so Comte *should* turn over in his grave. For the label *positivism* has been grossly misused by *both* advocates and critics. And worst of all, the term has also become a pejorative with which positivists are accused by critical theorists of being antihumanistic and "apologists for the status quo." These latter interpretations of Comte's vision would make him turn over again, for the very essence of his advocacy was to use laws of human organization to alter patterns of social organization for the better.

CONCLUSION

Thus, the Comtean message that was lost is this: Sociology can be a natural science; it can develop concepts that denote the basic properties of the social universe; it can develop abstract laws that enable us to understand the dynamics of the social universe; and if desired, these laws can be used to construct and reconstruct the social universe. This message is now considered naive in "sophisticated" theory circles. In my view, science is always naive in this fundamental sense. For if we assume that we cannot know something about the universe—that is, if everything is transitory, relative, value-laden, and subject to unresolvable "discourse"— then our newfound "sophistication" has taken us back into Comte's metaphysical stage.

The essays in the chapters to follow are by a positivist, who sees himself as working in the Comtean tradition. I cannot keep Comte from turning over in his grave, but I can stand him back on his feet. And I have chosen to do so by looking at classical theory through Comtean eyeglasses. While these essays were written at different times and with somewhat varying intents, they all reveal Comte's commitments to a science of human organization. Each essay on the early masters of sociology tries to extract the essential theoretical ideas from the thinker's work and to articulate the implicit laws and, in places, models proposed by a thinker.

Unlike most books on early sociological theory, including my previous ones, this one is short. There is a reason for this: Theory should be parsimonious; and so I move directly to the essence of each theorist's core ideas, ignoring context, biography, and other matters that theorists spend too much time pursuing. The goal of a science is to explain something, not to have endless discourse. And so the goal of these substantive essays is to get on with this task, to search out the explanatory principles. Most sociological theorists disagree with me on this score, preferring instead to debate the issues at a philosophical level. While these debates are intellectually interesting, they are also debilitating. They keep sociology from developing as a science.

The ultimate goal of a science is to incorporate the ideas of the early masters and relegate further discussion of them to other disciplines, such as history and philosophy. We should not write books on classical theory; we should formulate theoretical principles of basic and generic social processes. We should pull out the theoretically important ideas, formalize them, incorporate them with other formalisms, and move on, perhaps giving lavish credit in footnotes to the early founders. I do not think that this book on classical theory will eliminate books and discourse on early theory, but that is my goal.

NOTES

1. Comte was forced to abandon the term *social physics* because the Belgian statistician Adolphe Quetelet had usurped the term for a statistical approach—an approach much closer to the "mindless empiricism" later associated with Comte's term *positivistic* and, for that matter, with sociology in general.

2. For example, my colleague Randall Collins, whose work is certainly in the Comtean tradition, prefers not to use the label *positivist* because of its pejorative connotations and association with excessive concern with logic and deduction over insight and explanation.

2 Toward a Social Physics: Reducing Sociology's Theoretical Inhibitions

In the more developed sciences, it is generally recognized that the goal of "scientific activity" is ultimately to produce "theory." In the less developed sciences, such as sociology, this consensus does not exist. Indeed, there is disagreement over whether or not sociology can be a science, especially one modeled after the natural sciences. And even when scholars agree that sociology should be a science, there is debate over "what kind" of science it can be. Further, even with acknowledgment that the goal of scientific sociology is to generate theory, there is violent dissent over what theory is, what it should be, or what it can be.

To state the matter charitably, then, sociology is unsure of itself as a scientific discipline. It is uncertain about its goals. And when the goal is acknowledged to be "theory," one observes disagreement, controversy, and acrimony. In this chapter, I would like to offer a few observations on this unfortunate situation. First, I will review the diverse strategies in sociology for "building theory" and indicate how this diversity has ceased to be a source of intellectual vitality. Instead, it has become a cause of confusion and intellectual stagnation. Second, I will examine how these strategies reflect a series of intellectual inhibitions that sociological theorists have developed over the last fifty years. And finally, I will make an effort to show how these inhibitions can be overcome and how sociology can become a true science.

This chapter originally appeared in *The Humboldt Journal of Social Relations* 7, no. 1 (1979–80): 140–55.

As I address these issues, a persistent theme will be evident: We have lost the early vision of our first masters. In fact, we have become afraid to speculate about the important questions that consumed Comte, Spencer, Durkheim, Marx, and Mead. We have, in essence, traded in our armchairs for the comfort and security of statistical packages that, through the magic of computers, produce "results." Never mind that we all too frequently achieve statistical significance on insignificant matters. With a few notable exceptions, we have become overconcerned with a "subject matter that does not matter."

DIVERSE THEORETICAL STRATEGIES IN SOCIOLOGY

When examining theoretical strategies, we are really addressing several related issues. First, we are exploring "what form" theory should take when completed. That is, how should it be stated as a finished product? Second, we are asking "how does one go about constructing" theory. In other words, are there techniques, procedures, and rules for developing theory? This last question touches upon an issue that we cannot address: the creative act. Ultimately, a science builds upon the creative insights of its practitioners. As I will argue, theory begins with an insight into the properties of the universe. There is no form or formula for such insights; they simply occur. They are the product of genius and luck. But once they occur, these insights must be translated into a theoretical form; and it is in this translation that questions of theoretical strategy become paramount. For in the end, creative insights must be stated in ways that allow others to use them.

In sociology, there is a diversity of opinion on what constitutes insight and on the best way to express these insights. We can isolate at least five theoretical strategies in sociology: (1) the history of ideas strategy, (2) the concept formation strategy, (3) the empirical generalization approach, (4) the modeling procedure, and (5) the axiomatic approach. Each of these is viewed as theory in sociology by some group of practitioners. But rather than creating an exciting diversity of activity, this confusion over what theory is and should be has made sociology highly suspect in the scientific community. A brief examination of each strategy will, I feel, show why this should be so.

Theory as History of Ideas

A great deal of what is termed "theory" in sociology is tracing the history of ideas. Indeed, theory courses in universities are usually divided into "history of social thought" and "modern theory." Tracing just who influenced whom in the development of ideas is, of course, an exciting and important intellectual activity which justifies itself. But typically, the history of social thought does not produce theory; it involves the review of old ideas and their connection to historical conditions. Only in a discipline that is insecure, and in which theoretical activity has become recessive, could reanalyses of the early masters be considered theory. We need only imagine a physics text on theory examining the life and times of Michael Faraday to realize the absurdity of what sociologists do.

In arguing that the history of social thought is not theory per se, I am not asserting that Marx, Durkheim, Spencer, Mead, and others did not achieve fundamental insights. Indeed they did, but rarely does one see efforts among modern sociologists to extract key concepts and principles of the early masters. A developed science takes what is useful from its first masters and moves on. Sociology remains in a state of arrested intellectual infancy, afraid to leave the security of its early father figures.

Theory as Concept Formation

Concepts that denote basic properties of the universe are, of course, the cornerstone of all theory. To be theory, however, concepts must be related to each other in propositions. Unfortunately, this second step is frequently avoided. Much of what is called theory in sociology involves concept formulation and reformulation. Such activity is not improper per se, since making concepts more powerful and precise is an essential activity in any science. But all too often, theory in sociology becomes a game of definitions as the "real meaning" of a concept is asserted. Moreover, this fetishism over concepts is frequently mixed with theory as the history-of-social-thought strategy, as scholars debate what Marx, Weber, Durkheim, or Mead "really meant" by terms such as *alienation, self, verstehen, collective conscience*, etc. Modern sociological theory is often little more than arguments over definitions of such properties of the world as class, power, self, interaction, authority, influence, and the like.

The result of this definitional approach is that theory is arrested at a very preliminary stage: deciding what exists "out there" in the social world. Definitional disputes are best resolved as thinkers link concepts to each other and as researchers attempt to examine the empirical implications of such linkages. Redefinition will naturally occur as theorists and researchers "work with" concepts that are connected to each other, even if the definitions and linkages are provisional.

Theory as Empirical Generalizations

Another strategy for building theory in sociology revolves around constructing empirical generalizations about a particular phenomenon. For example, it is often considered "theoretical" to state that industrialization and family size are inversely related, that political democracy and modernization are positively associated, that capitalism and worker alienation are positively related, and so on. Such generalizations can become theoretical if efforts are made to abstract above them. For example, one might seek to discover how productivity, political centralization, and actor commitments in social systems are related. However, rarely is this necessary process of abstracting above the empirical content and context of generalizations initiated. Rather, the generalizations per se are considered theoretical, whereas in fact, they are merely regularities to be explained by abstract theories.

This confusion of *explanans* and *explanandum* is prevalent in sociology and leads us into many theoretical difficulties. If empirical generalizations are considered theoretical principles, then they will be time-bound to particular contexts and historical epochs. Moreover, they will be consistently "proven wrong," since they merely summarize modal tendencies in the empirical world at a particular point in time; exceptions will always be found in the past, present, or future. Thus, as long as sociologists confuse the thing to be explained with the theoretical explanation, theorizing will remain a frustrating activity. Additionally, it will encourage the current situation in sociology: the perpetuation of "theories of" substantive empirical areas. In sociology we can observe theories of such specific empirical phenomena as family, delinquency, class, modernization, ethnic groups, and bureaucracy. Such "theories" only state empirical regularities observed in these substantive areas of inquiry; they are the object of theoretical explanation, not the explanation.

Theory as a System of Categories

Much theory in sociology involves the elaboration of taxonomies. Classification can be a useful descriptive tool; it can allow us to "order the facts." Moreover, just as Linnean classification system stimulated Darwin, taxonomic systems can often encourage thinkers to speculate: Why do the "facts" order themselves in this way? An answer to such a question is likely to be theoretical. Yet modern sociology has come to view systems of categories as an end in themselves. A good deal of what is considered theory in sociology is little more than taxonomies of phenomena such as authority and power, types and stages of economic development, forms of organizations, types of communities, and other specific processes and entities in the social world. Some of this categorization is a variant of the "theory as concept formation" approach. But as Parsonian action theory illustrates (e.g., Parsons 1951), the taxonomic approach can involve elaborate systems of analytical categories designed to "order" specific empirical events. By finding "the place" of an event in the system of linked categories, advocates of this approach presume that the event in question has been explained.

Much like the "theory as empirical generalization" strategy, this approach confuses what is to be explained with the explanation. A system of categories that orders empirical observations is not an explanation any more than the "periodic table" in chemistry or the Linnean classification system constitute explanations. Taxonomies are descriptions of regularities in the universe that require a theoretical explanation. They are the beginning of theoretical activity, not its end.

Model Building as Theory

A model represents an effort to accentuate analytically the key properties, and their connections, of some phenomenon. As such, a model describes what is considered for some purpose to be the "important" properties and processes of a specific set of events or an entire class of events. While models often abstract above specific empirical cases and attempt to show how diverse empirical situations reveal certain common properties and processes, they are, nevertheless, a descriptive tool. Much like a category system, models allow us to determine the "place" of a specific event within a larger analytical configuration. But unlike taxonomies, models tend to be more dynamic, in that they emphasize (a) the connections among properties and events in the universe and (b) the capacity of one property or event to alter the nature or course of another.

In recent years, the dominant mode of theorizing in sociology has been the causal model. Basically, this strategy involves an effort to show how much variance in one event is "caused" by other events. The fact that this strategy involves an application of multivariate statistical analysis through the mechanical wizardry of the computer makes it highly appealing. But this fact also exposes such models for what they actually are: empirical descriptions of connections among specific empirical events in particular times and places. For example, causal models of occupational attainment are not theoretical, but descriptive. They typically describe the key variables in mobility patterns in American society, and perhaps Western societies, over the last few decades (e.g., Blau and Duncan 1967). They never reach a level of abstraction necessary to become theoretical. Moreover, efforts to do so would rob them of their basic utility: to trace causal connections among specific empirical variables. For as one moves up the abstraction ladder, one also pulls away from the specifics of causality in a particular situation, thus rendering the causal model less and less useful.

Thus, sociologists often confuse, once again, the thing to be explained—in this case, patterns of causal connections among empirical variables in particular contexts—with theoretical explanation. For instance, occupational mobility patterns in America are but one type of more generic phenomena such as movement of individuals in social hierarchies. Principles of such movement, should they be developed, will constitute the theoretical explanation of occupational mobility in America and elsewhere. Thus, the causal model is simply a refined and highly useful form of description, but it is not, and cannot be, a theory.

Articulation of Principles as Theory

In the more developed sciences, explanation occurs in terms of the application of an abstract principle to a particular set of empirical events. Such principles state at a highly abstract level the fundamental relations of generic properties of the universe and are then used, sometimes in conjunction with other principles, to understand why the universe operates in certain ways. This type of explanation is labeled "axiomatic" when precise and logical deductions from abstract principles to particular empirical hypotheses are performed.

In recent years, a number of sociologists (e.g., Homans 1961; Blau 1977; Emerson 1972) have sought to employ this form of explanation. Sometimes such efforts involve such rigid concern with the logic of deduction that they lose the capacity to deal with interesting and important questions. Yet there are a number of useful efforts at generating sociology's first principles to argue for the power of this approach. Indeed, it was the approach employed by sociology's first masters, particularly Comte, Spencer, Marx, and Durkheim. Underlying all of their efforts was an attempt (a) to discover the fundamental properties, and their relations, of the social universe, and then, (b) to express these properties as a set of more abstract principles. Some of these early masters, such as Spencer (1876), did this explicitly, whereas others, like Mead (1934) and Durkheim (1893), did so implicitly.

If sociological theory is to recapture the excitement of its first theorists—an excitement that we still sense in reading them—it must reorient itself to the search for the basic principles that will allow us to understand the social universe. Such a search should not become sterile through overconcern with the form of axiomatic deduction. Rather, it must be guided by several general assumptions: (1) Behind the surface variations and varieties of empirical forms that typify the social world, there are certain common properties of all human interaction and organization. (2) These properties can be expressed in a series of abstract principles. (3) Such principles will become equivalent to those in physics in that they will form the core of our understanding of the social world. (4) There will be comparatively few principles of this nature.

Without the conviction that laws of human organization can be discovered, sociological theory will be arrested. Moreover, we will continue to summarize the masters, define and redefine concepts, confuse empirical generalization with explanation, elaborate taxonomies, and believe that time-bound causal models of empirical events are theory. Unfortunately, there are a number of intellectual and professional barriers to redirecting sociology. I would now like to examine the most important of these.

SOME THEORETICAL INHIBITIONS (OR EXCUSES FOR NOT BUILDING THEORY)

One of the most frequently cited reasons for the lack of mature theory is that sociology is a young science. This is an excuse, since the discipline is 150 years old and since humans have been offering speculations on the human condition for thousands of years. As long as sociology remains comfortable with this attitude, it will not develop into a mature science.

Another explanation for sociology's theoretical failings comes from those who argue that sociology has an insufficient data base for either inducing theory or testing its implications. This position follows from a false conception of theory. In actual practice, theory is often generated without extensive knowledge of "the facts." Indeed, empiricists are rarely able to abstract above "the facts"; formulating concepts and formulating theories require different mental processes than does data gathering. Thus, if sociology waits for the accumulation of more "facts," it will continue to inspire new data analysis techniques, but it will thwart the development of scholars with the capacity to develop theoretical principles that can organize research.

Still another explanation for sociology's failings is presented by those, such as Merton (1968), who argue for middle-range theories as the necessary prerequisite for more general theoretical principles. This advocacy has dominated sociology for the last two decades, but unfortunately, the middle-range strategy has not been implemented in the way Merton intended. Rather than theories of limited range—in terms of their levels of abstraction and breadth of coverage—we have generated a series of highly specific theories in a number of diverse substantive areas that are, in many ways, little more than collections of generalizations about empirical findings. Instead of well-developed theories on generic social processes and on basic types of social structures, sociology is a collection of "theories" of such substantive phenomena as family, criminal gangs, finance units in organizations, economic development, ethnic minorities, and the like. These are not middle-range theories; they are places where one might test a theory. To paraphrase George Homans (1961), they are "where one studies, not what one studies." The result of the middle-range strategy has created a series of interesting empirical generalizations, typically presented as a causal model or some other correlational device, *as if* they were theory. As I emphasized earlier, sociologists often confuse the procedures for testing theories with the process of constructing theory.

Yet another explanation for sociology's lack of accepted theory comes from those who view the "natural science" conception of theory as impossible in analyzing the social world. There are those who believe that human behavior and organization contain the capacity for spontaneity and indeterminism; and hence, that there are no timeless, or universal, patterns of organization that are describable in terms of abstract laws (Blumer 1969). Others argue that each historical epoch reveals its own laws of organization, rendering the search for panhistorical, or universal, laws fruitless (Appelbaum 1978). Another group of scholars believes that the methodological problems of humans studying humans are so great as to make deductive theory and its definitive refutation a virtual impossibility. Or, at the very least, that facts require as a first priority the discovery of the laws of human thought, cognition, and consciousness, since all knowledge about patterns of social organization is mediated through such mental processes (Cicourel 1964, 1973).

These explanations and their variants arrest our theoretical imagination. They tell us that abstract theory cannot be developed for social phenomena; or if it can, then it must wait for a more adequate data base, a body of middle-range theories, and worst of all, an unspecified number of years until intellectual maturity sets in. These explanations are feeble excuses, and they are incorrect. Theory in other sciences has often come early in the discipline's history; it has frequently come without extensive catalogues of facts; and it has had to overcome methodological obstacles equal to those in the social sciences. (We could view the fact that we are humans studying humans as an *advantage* rather than a handicap, since we can intuitively achieve familiarity with our data.)

REMOVING SOCIOLOGY'S THEORETICAL INHIBITIONS

Once we refuse to accept the above excuses for sociology's theoretical failings and once we commit ourselves to search for general and abstract principles, we have moved closer to Auguste Comte's (1830–1842) original vision of "social physics." The goal of theory is no longer one of tracing "causality," "explaining variance," and controlling for "extraneous variables." (These are the concerns of researchers.) It no longer involves disguising empirical generalizations as "theories of." And it reduces our current mania for concept fetishism and taxonomy. Instead, we move up the abstraction ladder and attempt to understand the nature of events by isolating their key properties and relations.

Such a shift in emphasis involves restating typical sociological questions in the abstract language of theory rather than the contextual concerns of researchers. For example, researchers are interested in marital interaction, while theorists are concerned with patterns of prolonged interaction in intimate contexts, whether in a marriage or in some other empirical situation; researchers are likely to study gang delinquency, while theorists are concerned with nonnormative behavior; researchers may be fascinated by the causes of educational achievement, whereas theorists are interested in the properties of movement of actors in hierarchies; researchers are concerned with economic development, whereas theorists are attuned to basic patterns of productivity in human systems; researchers may wish to trace the history of dominance by nation-states during capitalism, while theorists are concerned with dominance relations per se; and so on.

Perhaps this shift in emphasis seems obvious, but one can find a "theory of" each research area cited. This fear of abstraction is probably the most important barrier to sociological theory. The incessant worry over whether or not a "theory is testable," whether or not concepts have been "operationalized," and whether or not "research and theory have been linked" all operate to arrest our imaginations. These are the concerns of researchers, not theorists. If an abstract proposition strikes us as useful and insightful, ways will be found by researchers to operationalize its concepts, develop an appropriate research methodology, and use it to interpret existing research findings. To worry about these issues when seeking to build theory allows existing research protocols to dictate our intellectual problems and to dominate our thought processes.

TOWARD A SOCIAL PHYSICS

The Proper Orientation

The term *physics* was not originally identified with a particular natural science. Indeed, it originally denoted the search for the "nature of" phenomena in the universe. Comte's vision was to develop a science that searches for the nature of social phenomena in the universe. This vision still holds good for sociologists, since theory must begin with some conception of the key properties evident in all social systems. Theory searches for the generic, not for the specific, time-bound, or contextual. Thus, a return to "social physics" begins with an effort to isolate properties of the social world that transcend particular times, places, and contexts. It begins with a concern about what is basic to all patterns of social organization.

Some of the early masters clearly understood this primary goal of theory. In looking at their works, we see a concern with universal properties of the social universe: interaction, socialization, productivity, differentation, integration, and so forth. Their formulations were not about specific events, but about what they saw as a fundamental property of the social world. They did not develop theories of _____ (fill in the substantive area); they theorized about the nature of phenomena and sought to develop concepts that captured the essential properties of these phenomena.

The next step in social physics is to specify the key relationships among basic properties of the universe. This is not done by correlating empirically defined variables; such efforts are not sufficiently abstract to be of great use to theorists. Rather, it is done through a "creative act" on the part of the theorist. Familiarity with existing data can help, of course; there can be little doubt about this. But ultimately, as Willer and Webster (1970) have pointed out, theorists engage in a process of "abduction"—of employing simultaneously the logic of induction and of deduction. The key ingredient in making connections among properties of the universe is to leave one's imagination open, to be uninhibited by the partial correlations of empiricists, to use our own experiences and familiarity with the world as we know it, and to seek constantly to abstract above details, contexts, and the specifics of particular times, places, and situations. As long as we believe that all events are contextual, that people's actions are indeterminate, that historical epochs reveal their own unique logic and dynamics, that data collection must precede theorizing, that we are too young a discipline to think this way, that we have not controlled for some set of contextual variables, and so on, then we will never become a true science.

For in the end, theory is abstract; it seeks to simplify relationships by rising above empirical particulars and by showing the underlying properties and dynamics of these particulars. Theory also assumes that these underlying properties do not change, that only their concrete manifestations vary under different empirical conditions. Until we approach our subject matter with this orientation, social physics will not be possible. And we will continue to wallow in our theoretical inhibitions.

3 The Forgotten Theoretical Giant: Herbert Spencer

Unlike the works of other intellectual giants of the nineteenth century, the works of Herbert Spencer have received relatively little attention. At the same time that Marx, Weber, Durkheim, and Simmel are revered, Spencer is forgotten and often intellectually vilified. Indeed, if we were to construct a portrayal of Spencer, or an intellectual obituary, it might read as follows (Turner and Beeghley 1978):

> Herbert Spencer, the first self-conscious English sociologist, advocated a sociological perspective that supported the dominant political ideology of free trade and enterprise. He naively assumed that "society was like an organism" and developed a sociology that saw each institution as having its "function" in the "body social"—thereby propagating a conservation ideology and legitimating the status-quo. Anticipating Darwin's concept of "Natural Selection," Spencer coined the phrase, "survival of the fittest," to describe the normal state of relations within and between societies—thus making it seem right that the elite of a society should possess privilege and that some societies should conquer others.

Against such a view, Talcott Parsons's (1937, 1) early observation—"Who now reads Herbert Spencer?"—may seen appropriate. But unfortunately, not only was Parsons's query an accurate description for the 1930s, it was also to be an accurate prophecy for the next four decades. Today, few read Spencer; and among those who do comment on his ideas, numerous misinterpretations prevail.

This chapter originally appeared in *Revue européenne des sciences sociales* 19(1981): 79–98.

This chapter seeks to examine Spencer as a sociological theorist, rather than as a moralist, philosopher, or political commentator.[1] In so doing, I hope that Spencer's purely sociological works will be better appreciated by an intellectual community that has too readily dismissed this intellectual giant of our past. In focusing only on Spencer's theoretical models and principles, I am ignoring many of his most important contributions to methodology (Spencer 1873) and to ethnographic disciplines (Spencer 1873–1934), but hopefully, a renewed appreciation of Spencer's theoretical models and principles will encourage a less prejudiced reading of his other works.

SPENCER'S THEORETICAL MODELS

In his *First Principles*[2] (1862, 3, 43), Spencer defined cosmic evolution in the universe as an "integration of matter and concomitant dissipation of motion; during which the matter passes from an indefinite incoherent homogeneity to a definite coherent heterogeneity; and during which the retained motion undergoes a parallel transformation." In this law or "first principle," Spencer felt that he had unlocked one of the common properties of the inorganic, organic, and superorganic (social) realms. Indeed, Spencer argued that this general principle applied to the evolution of solar systems, organic life, and social systems. While such an assertion is, no doubt, untrue, Spencer's vision of social systems was always conditioned by this general view of evolution. But immediately, we must add several important qualifications to this often-made observation (Peel 1972; Carneiro 1967).

First, when applying this law to social systems, Spencer devoted his attention to the process of institutionalization (Perrin 1976). For although Spencer was concerned with long-run evolutionary trends in societies, he also used the concept of evolution to denote short-term processes of "structuring" in social systems. Secondly, this emphasis is underscored by the fact that in formulating this law, Spencer (1862) devoted many pages to the analysis of "dissolution" in which structures "dissipate" and move toward "incoherent homogeneity." Thus, when Spencer employed the term *evolution* and applied it to social systems, he was addressing the twin processes of institutionalization and deinstitutionalization, or structuring and unstructuring of social relations.

Too often, Spencer is viewed as a naive evolutionist who saw societies as marching toward the Western ideal of industrial capitalism. Such assertions represent a profound misreading and unfair evaluation of Spencer's sociological works. This tendency to misread Spencer can be demonstrated by articulating Spencer's models of social organization. When Spencer's three analytical models are examined, only one of the three is explicitly evolutionary, and it is far from naive or simplistic. His other two models are abstract and analytic in that they focus on the general process of institutionalization and on the underlying dynamics of such institutionalization. Thus, to appreciate the power of Spencer's analysis, we need to examine his three basic models on (1) the process of institutionalization, (2) the phases of institutionalization, and (3) the process of societal evolution.

The Process of Institutionalization

For Spencer the process of institutionalization involves growth in the size of a population, its differentiation, its integration, and finally, its adaptive upgrading.[3] When viewed in these terms, much of the awkward terminology in Spencer's definition of evolution is clarified. Institutionalization involves the aggregation of "matter" (individuals and social units), which, by virtue of being brought together, evidence "motion" or tendencies to act in certain ways. This "retained motion" leads to differentiation of actions, and as individuals and social units become segregated in space, their propensities to act differently become amplified and multiplied over time (from "homogeneity to heterogeneity"). But at some point the "motion" that pushes units in different directions becomes "dissipated" as the units encounter resistance to their actions. Moreover, if the aggregated units are to be kept from dispersing, they must become "integrated" through the centralization of authority and mutual interdependence. And as these processes of integration occur, the aggregated units become more "coherent," resulting in an increased level of adaptive capacity to their environment. At any point in this process (Spencer 1862), dissolution can occur, if the retained motion is stronger than the forces causing aggregation, if segregation and multiplication of differences are too great, if centralization of authority is insufficient, and if mutual interdependence cannot be established. Spencer's intent can be illustrated with a hypothetical example: A society that grows as the result of conquering another will tend to differentiate along conquered and conqueror lines; it will centralize authority; it will create relations of interdependence; and hence it will become more adapted to its environment. The result of this integration and adaptation is an increased capacity to conquer more societies—hence, setting into motion another wave of growth, differentiation, integration, and adaptation. At any point in this process, however, dissolution can occur, if the retained motion of the conquered is greater than the force of their conquerers, if centralization of authority cannot become legitimated and effective, and if mutual dependence cannot be established. Similarly, Spencer's ideas can be applied to nonsocietal social systems, such as corporations, which, for example, might begin to grow through mergers or expenditures of capital. But soon, Spencer would have argued, it must differentiate functions and then integrate them through a combination of mutual dependence of parts and centralization of authority. If such integration is successful, it has increased the adaptive capacity of the system, and it can grow, if some "force" (such as capital surplus) is available. Thus, Spencer's "law of cosmic evolution" provides a broad metaphor for specifying certain key processes creating and elaborating social structures: (a) forces causing growth in system size (whether by compounding smaller units or by internal creation of new units); (b) the differentiation of units in terms of segregation and multiplication of effects (the "homogeneity" to "heterogeneity" portion of the law of evolution); (c) the processes whereby differentiated units become integrated (the "integration of matter" and "dissipation of motion" portions); and (d) the creation of a "coherent heterogeneity" that increases the level of adaptation to the environment. This implicit model is made explicit in figure 3.1.

Figure 3.1 outlines the stages of institutionalization. As is emphasized, the fundamental processes of growth, differentiation, integration, and adaptive upgrading are, to some extent, conditioned (a) by "external factors," such as the availability of natural resources; (b) by "internal factors"—e.g., the nature of the internal units; and (c) by "derived factors," such as the existence of other societies or internal values and beliefs.[4] And as is also evident, Spencer's definition of evolution is rendered more understandable. Some "force," whether economic capital, a new technology, a need to gather resources, new values and beliefs, etc., sets into "motion" system growth. This "motion," as it acts differently on various units, sends them in different directions and "segregates" them, such that their differences are "multiplied" as the retained motion allows for their elaboration. Yet, if the system is not to explode, the units of "matter" must be "integrated," thereby dissipating or channeling the motion of the parts in ways that increase the "coherence" of the whole. Such coherence increases the adaptive capacity of the system. Conversely, to the extent that integration is incomplete and/or the force that drives the system is spent and cannot be replaced, dissolution of the system is likely. Thus, social systems grow, differentiate, integrate, and achieve some level of adaptation to the environment, but at some point their driving force is spent or units cannot become integrated, setting the system into a phase of dissolution.

The Model of System Phases

Spencer visualized that during institutionalization, social systems cycle through phases in which authority becomes highly centralized, and

FIGURE 3.1. Spencer's Model of Institutionalization

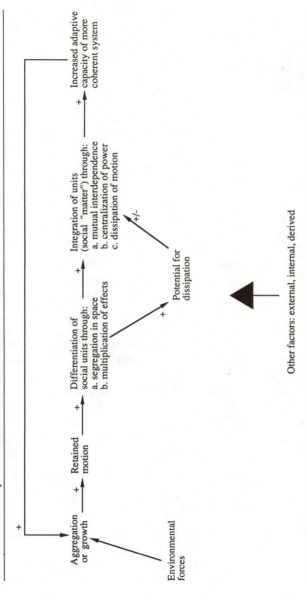

then less centralized. Spencer saw centralization as primarily a response to environmental threats and conflict or to extreme internal diversity. But he also recognized that the processes of differentiation and integration, per se, are involved (Spencer 1874–1896). Social systems will be centralized rapidly under conditions of external threat, but regardless of these forces, they possess an inherent dialectic, in that decentralized systems experience pressures for increased centralization, while highly centralized systems reveal pressures for less direct control by centralized authority.

Contrary to much of the commentary on Spencer (e.g., Sorokin 1961; Coser 1977), it is this cyclical process that Spencer denoted with his distinction between "militaristic" and "industrial" societies. Too often, this distinction is viewed as an evolutionary sequence, but such a view represents one of the most consistent misreadings of Spencer's *Principles of Sociology* (1876, vol. 1:449–597). For Spencer, *militaristic* denotes the degree of centralization of power and control of internal system processes, whereas *industrial*[5] refers not to a particular mode of economic production (such as industrial capitalism) but to the deregulation of internal system processes. Hence both traditional and modern societies can be either *militant* or *industrial*—a point of emphasis that is ignored in efforts to view Spencer's distinction between *militant* and *industrial* as denoting a unievolutionary trend.[6] When this fact is recognized, a model such as that diagrammed in figure 3.2 emerges.

If we begin analysis of the phases, as Spencer would have intended, the cycle is initiated with differentiation and diversification of the system and its constituent parts. Such growth leads to integrative problems which increase with further diversifications of units, eventually creating pressures for consolidation of differentiated and diversified units. At some point, and under variable empirical conditions, these pressures lead to centralization of authority, which results in tight control of internal operative and distributive processes by regulatory centers. Over time, such control creates stagnation by limiting the developmental options of system units, with the result that pressures for deregulation mount. At some point under varying empirical conditions, these pressures lead to decentralization, which sets off a new wave of differentiation and diversification.

This model, which appears in *Principles of Sociology* (1876, 576–87), supplements Spencer's general view of institutionalization by specifying the more rhythmic cycles that occur during institutionalization. In turn, both the general model of institutionalization and this model on the phases of institutionalization provide some insight into Spencer's view of the dynamics underlying long-term evolutionary development.

FIGURE 3.2.   Spencer's Model of System Phases

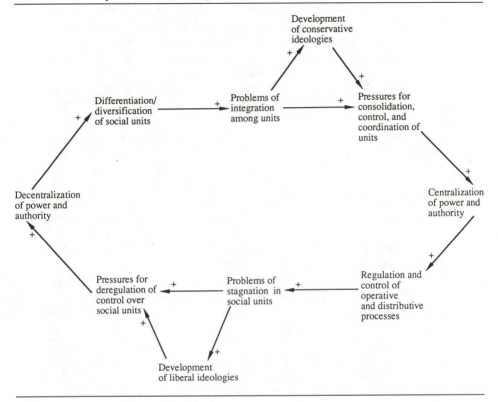

The Model of Long-term Societal Evolution

Curiously, Spencer's long-term evolutionary model is rarely discussed in commentaries on his work (indeed, the militant-industrial distinction is discussed as if it were the long-term evolutionary model). But Spencer's descriptive model of evolution is far more sophisticated than is usually realized and is, in fact, equal to those developed in recent times (Parsons 1966, 1971; Lenski 1966, 1970; Turner 1972).

As Spencer's definition of evolution underscores, human evolution involves movement from simple and homogeneous societies to increasingly more differentiated and complex systems. As such, societal evolution is but one type of more general evolutionary processes in the cosmos. Societal differentiation, Spencer felt, occurs along three broad classes of functions[7]—regulatory, operative, and distributive—and as such, societal evolution reveals a parallel to growth and development in organic bodies.[8] The first differentiation is between the regulatory and operative, but with growth of, and differentiation within, structures performing these functions, separate distributive structures emerge. Subsequent evolution involves growth of, and internal differentiation within, each of these three classes of structures. Thus, for example, the regulatory system of a society will initially differentiate into separate administrative (for internal affairs) and military (for external relations) subsystems. And with further evolution, the military and administrative branches grow and differentiate internally, while a new type of regulatory structure, the monetary, differentiates from the military and administrative. Similarly, operative and distributive structures become increasingly differentiated from the regulatory and from each other, while becoming internally differentiated.

Spencer labeled four conspicuous stages in this evolutionary process: (1) simple,[9] (2) compound, (3) doubly compound, and (4) trebly compound.

Simple societies are those where regulatory, operative, and distributive processes are not greatly differentiated. Compound societies are created by the joining of simple societies or by internal growth. They reveal clear differentiation between regulatory and operative processes, as well as some internal differentiation within each. However, distributive structures are not clearly separated from either regulatory or operative processes. But in double compound societies, where separate regulatory and operative systems have undergone further growth and internal differentiation, a separate set of distributive processes does become clearly differentiated. And in trebly compound systems, regulatory, operative, and distributive structures expand and differentiate even further.

This pattern of long-run growth and differentiation in human societies is represented in figure 3.3. This model reproduces in diagrammatic form much of information provided in the later volumes of *Principles of Sociology* (volumes 2 and 3).[10] This vision of evolution is inspired by a biological model of growth in organisms (Spencer 1862, 1864–1867), since societal evolution involves growth and differentiation along three major functional systems that have their analogues in more complex animal organisms. Such analogizing can be viewed as a limiting constraint, but it allowed Spencer to capture many of the most salient properties of long-term societal evolution. Indeed, the model is far more sophisticated than those developed by any of Spencer's contemporaries, such as Marx, Comte, Weber, Durkheim, Tonnies, Maine, or Tyler; and it is equal to those models developed by contemporary sociologists and anthropologists.

In sum, then, these three models reveal considerable sophistication in Spencer's sociological analysis. For whatever his politics and moral philosophy,[11] his sociological analysis captures many of the important dynamics of social systems. Yet despite their heuristic value, these models provided the assumptive base for Spencer's most important theoretical contribution: the development of some abstract laws or principles of social organization.

SPENCER'S THEORETICAL PRINCIPLES

With the exception of his first principles of evolution, Spencer never formally stated his principles of sociology. Yet he recognized that as a science, sociology requires the formulation of abstract principles that denote fundamental relationships among phenomena in the social, or what he called the "super-organic," realm. While these principles would be connected to the universal principles of evolution, and while they would reveal some affinity to the principles of biology or organic systems, they would, nevertheless, be unique to sociology, since social systems constitute a distinctive realm in the universe. And thus, while he did not state his sociological principles formally, nor as abstractly as I intend to state them, it is relatively easy to extract abstract principles from his more discursive arguments. For present purposes, these principles can be organized under three general headings: (1) principles of growth and differentiation, (2) principles of internal differentiation, and (3) principles of differentiation and adaptation.[12]

Principles of Growth and Differentiation

Spencer saw a fundamental relationship in the social universe between the size of a social aggregate and the process of structural differentiation.[13] He couched this insight in the metaphor of "growth," with the result that differentiation in social systems is a positive function of increases in the size of a social aggregate. And with this growth metaphor, he specified additional relationships: The rate of growth and the degree of concentration of aggregate members during growth are also related to structural differentiation. Moreover, he saw that increases in system size are, to a very great extent, related to previous increases in the size and level of differentiation of an aggregate. Those aggregates that are able to integrate differentiating social units at one point in time are in a better position to increase their size and level of differentiation at a subsequent point. These insights into the relationships among growth, size, and differentiation in social systems can be expressed in the following propositions (Spencer 1864; 1875, 471–90):

1. The larger a social system is, the greater its level of structural differentiation.

FIGURE 3.3. Spencer's Stage Model of Evolutionary Differentiation

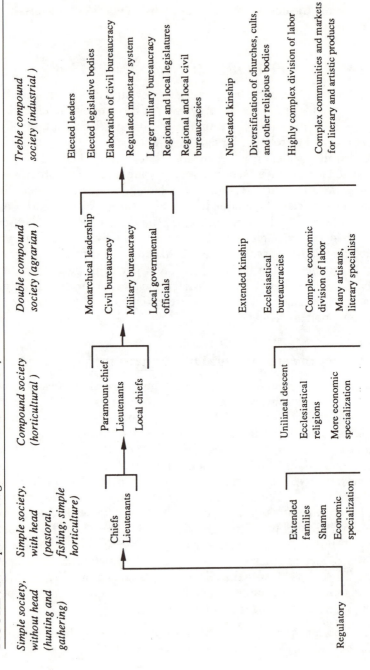

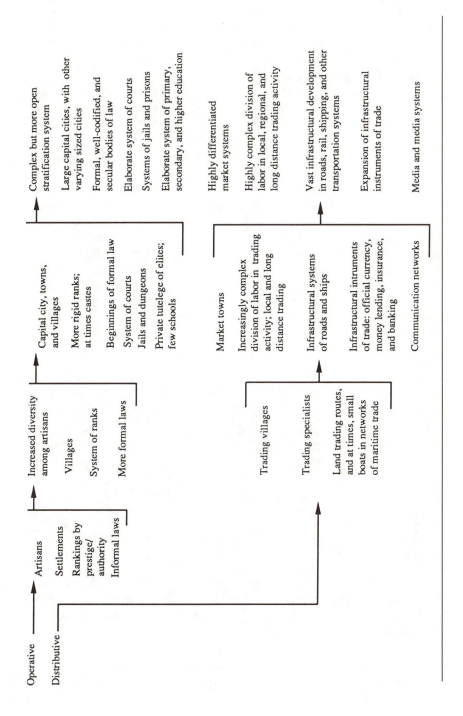

Operative ⟶ Artisans ⟶ Increased diversity ⟶ Capital city, towns, ⟶ Complex but more open
Distributive   among artisans   and villages   stratification system

                  Settlements   Villages   Large capital cities, with other
                                                                 varying sized cities

                  Rankings by   System of ranks   More rigid ranks;   Formal, well-codified, and
                  prestige/   at times castes   secular bodies of law
                  authority

                  Informal laws   More formal laws   Beginnings of formal law   Elaborate system of courts

                                                           System of courts   Systems of jails and prisons

                                                           Jails and dungeons   Elaborate system of primary,
                                                                                        secondary, and higher education

                                                           Private tutelege of elites;
                                                           few schools

                                     Market towns   Highly differentiated
                                                                                 market systems

                  Trading villages   Increasingly complex   Highly complex division of
                                     division of labor in trading   labor in local, regional, and
                                     activity; local and long   long distance trading activity
                                     distance trading

                  Trading specialists   Infrastructural systems   Vast infrastructural development
                                        of roads and ships   in roads, rail, shipping, and other
                                                                                  transportation systems

                  Land trading routes,   Infrastructural intruments   Expansion of infrastructural
                  and at times, small   of trade: official currency,   instruments of trade
                  boats in networks   money lending, insurance,
                  of maritime trade   and banking

                                     Communication networks   Media and media systems

2. The greater the rate of growth of a social system, the greater its rate and degree of structural differentiation.

3. The more the growth in the number of members in a social system is concentrated, the more likely that growth is to be accompanied by high rates of structural differentiation.

4. The more that growth and differentiation at one point in time result in structural integration of system units, the more likely that system is to grow and differentiate further at a subsequent point in time.

Principles 1 and 2 state what Spencer felt to be invariant relations, regardless of other conditions. Principles 3 and 4 are expressed more probabilistically, since specific empirical conditions can influence the likelihood that differentiation will follow from concentration alone and from previous patterns of integration. But just whether these principles represent invariant laws or probabilities is probably less important than Spencer's basic insight: In the social world, growth, size, and differentiation are fundamentally related. Spencer was the first to clearly perceive this basic relationship, and it marks one of his more enduring contributions to sociological theory. Indeed, these principles have found expression in a number of contemporary contexts, such as human ecology (Hawley 1950), organizations theory (Blau 1970), and analyses of states (Nolan 1979).

Principles of Internal Differentiation

Spencer did more than attempt to relate structural differentiation in social systems to size, growth, population concentration, and previous patterns of integration. He also sought to understand the nature and course of structural differentiation in social systems.[14] The first of these principles concerns the sequence of differentiation among different types of structures in social systems (Spencer 1875; 1876: 491–548).

1. The more a social system has initiated the process of structural differentiation, the more likely that the initial axis of differentiation will lie between regulatory and operative structures.

2. The more a social system has differentiated separate regulatory and operative structures, and the greater the volume of activity in that system, the more likely that separate mediating structures involved in distributive processes will become differentiated from regulatory and operative structures.

Once separate regulatory, operative, and distributive structures become differentiated, there are pressures for integration. Spencer saw two major mechanisms of integration:[15] (a) mutual dependence of diverse units on one another and (b) centralization of authority in the regulatory system. Hence, Spencer offered the following principle:

3. The more differentiated the three major axes in a social system, the greater its integrative problems, and hence, the more likely that relations of mutual interdependence and centralized authority will develop in that system.

For Spencer, then, differentiation in social systems occurs along three general axes, occurs in a particular order, and eventually generates pressures for integration. Subsequent differentiation among either the regulatory, the operative, or the distributive structures also occurs, Spencer argued, in a particular pattern and sequence. The pattern of differentiation for the regulatory, operative, and distributive systems can thus be expressed as follows (Spencer 1862, 1864–1867, 1875: vol. 7, 1874–1896, 491–548).

4. The greater the degree of differentiation along the regulatory axes, the more likely it is that differentiation will initially occur between structures dealing with (a) the external environment and (b) internal activities, and that only after the differentiation of (a) and (b) is differentiation of regulatory structures for facilitating the exchange of resources likely to occur (Spencer 1864; 1875, vol. 6).

5. The greater the degree of differentiation along the operative axes, the more likely are diverse activities to become spatially separated and localized (Spencer 1862; 1875, vol. 6: 491–97).

6. The greater the degree of differentiation along the distributive axes, (a) the greater is the *rate* of movement of materials and information in the system, (b) the greater is the *variety* and volume of materials and information distributed in the system, and (c) the higher is the *ratio of information to materials* distributed in the system (1876, 509–12).[16]

These propositions are stated more abstractly than intended by Spencer, who was primarily concerned with societal social systems. Yet they follow Spencer's intent, which was to stress that as the regulatory system differentiates, it initially creates separate structures for dealing with internal and external problems, such as a military and civil bureaucracy in a societal system. Later, structures evolve to facilitate the flow of materials and other vital resources, since regulation of large and complex systems is possible only with common distributive media, such as money in societal systems. These insights, expressed in proposition 4 above, also apply to other types of social systems, such as organizations, communities, and perhaps large groups. Proposition 5 states that differentiation, in accordance with the laws of force, motion, segregation, and multiplication of effects (Spencer 1862), tends to create ever more diverse structures, with similar activities becoming spatially concentrated into districts that are distinguishable from each other in terms of their types of activity. Proposition 6 stresses that as distributive structures differentiate, there is greater need for increased speed and volume in the movement of materials and information, since the expansion of distributive structures is a direct response to growth in the size and complexity of operative and regulatory processes (Spencer 1864–1867; 1874, 5, 6).

Spencer recognized that the nature of internal differentiation varied in terms of external and internal conditions. Systems in one kind of external environment, or with a particular composition of units, would become internally structured in ways that would distinguish them from systems in different external and internal circumstances. Spencer saw war and diversity of races as the key external and internal variables. And if we abstract Spencer's above discussion of these variables for societal systems, the following principles become evident (Spencer 1876, 519–48):

7. The greater the degree of external environmental threat to a differentiating system, the greater the degree of internal control exercised by the regulatory system.

8. The greater the degree of threat to system stability posed by dissimilar units, the greater the degree of internal control exercised by the regulatory system.

9. The greater the degree of control by the regulatory system, the more growth and differentiation of operative and distributive structures are circumscribed by the narrow goals of the regulatory system.

Spencer recognized, however, that both external and internal control create pressures for their relaxation (see figure 3.2). And with less regulatory control, internal diversity and dissimilarity increase, with the result that eventually pressures would be created favoring centralization of regulatory control. Thus, two final propositions can be extracted from Spencer's analysis in *Principles of Sociology* of internal differentiation in social systems (Spencer 1876, 576–97).

10. The more operative and regulatory structures are circumscribed by centralized regulatory structures, the more likely they are, over time, to resist such control, and the more they resist, the more likely is control to decrease.

11. The less operative and distributive processes are circumscribed by centralized regulatory structures, the greater are problems of internal integration, and the more likely is the regulatory system to increase efforts at centralized control.

Principles of Differentiation and Adaptation

Spencer felt that differentiated social structures are better able to adapt to environmental conditions.[17] As he stated in *First Principles*, homogeneous masses are unstable and vulnerable to disruption by environmental forces, whereas differentiated systems, with their relations of interdependence and centralized regulatory apparatus, can cope more effectively with the environment, frequently using it for their purposes. This view of differentiation and adaptation can be expressed as follows:

12. The greater the degree of structural differentiation in a system, and the greater its level of internal integration, the greater will be its adaptive capacity.

And as emphasized in proposition 4, superorganic systems that are well adapted to their environments are capable of further growth and differentiation. Well-organized systems can extract resources from the environment, or they can join, absorb, or conquer other systems in their environment.

CONCLUSION

The purpose of this chapter has been twofold: (1) to correct some of the consistent misinterpretations of Spencer and (2) to highlight his theoretical models and principles. When this is done, Spencer's work reveals far more insight and sophistication than is typically acknowledged. Indeed, Spencer's principles of differentiation and integration may represent some of sociology's "first principles" or abstract laws. And it should be emphasized that these principles were formulated almost twenty years before the appearance of Durkheim's ([1893] 1933) analysis of the division of labor, which repeats many of Spencer's ideas.[18] And Spencer's principles appeared from fifty to almost one hundred years before current work on such topics as size and differentiation in complex organizations (Meyer 1972; Blau 1970; James and Finner 1975), growth and administrative intensity in organizations (Hage, Aiken, and Marrett 1971), patterns of ecological growth and segmentation in communities (Hawley 1950; Stephan 1971), differentiation and concentrations of power in nation states (Szymanski 1973; Rueschemeyer 1977), size and administrative intensity in nations (Nolan 1979), population and ecological effects on structural differentiation in organizations (Hannan and Freeman 1977), and size and interaction densities (Mayhew and Levinger 1976). While these and other works in divers contexts expand upon Spencer's ideas, Spencer's principles are, in many ways, the abstract laws from which many of these sociological propositions on growth, size, ecological distribution, differentiation, and integration can be deduced.

Some of the work in these contexts acknowledges Spencer's contribution, but much appears to have been conducted without knowledge of Spencer's principles. Successive "new discoveries" of old principles is, however, an inefficient way to build sociological knowledge. Moreover, by inducing Spencer's principles from diverse empirical contexts, the common properties of these contexts are often ignored or not seen. With a more deductive emphasis than is typical in sociological research, in which Spencer's principles are initially used as laws and from which deductions to diverse empirical settings are made, interaction between research and theory could more readily occur. But as long as the abstract principles of a scholar like Spencer are ignored, the general laws of sociology will not be tested, and research will continue to partition theoretical inquiry and confine theoretical speculation to different empirical contexts, whether "organizations theory," "ecological theory," "political economy," or other areas of specialization in sociology. Thus, the Spencerian principles listed in this chapter are believed to still be plausible and useful. Reformulation at an abstract level, and testing at an empirical level, are therefore invited.

I should emphasize again that Spencerian principles have been used for decades in a wide variety of empirical contexts. Indeed, we could venture that they have been used in empirical research far more often than principles developed by Marx, Weber, and Durkheim. Sometimes this usage is acknowledged, but more often it is unknown, with the result that Spencer's ideas have often had to be rediscovered. We can further speculate that had sociological theorists and researchers begun the twentieth century with Spencer's models and principles in hand, it is likely that sociology would now be a more mature science. Hopefully, an exercise like that performed in this chapter can not only rekindle interest in this forgotten giant, but can also serve as a theoretical stimulus in the same way that Marx, Durkheim, and Mead still inspire theoretical and research activity.

NOTES

1. Spencer's philosophic, political, and moral works. Most of the prejudices against Spencer stem from his first major work, *Social Statics* (1850), and his last, *Principles of Ethics* (1892–1898). But his major nonsociological works, *Principles of Psychology* (1855), *Principles of Biology* (1864–1867), and *First Principles* (1862) are virtually devoid of political and moral commentary. And his major sociological works, *The Study of Sociology* (1873), *Principles of Sociology* (1874–1896), and *Descriptive Soci-*

ology (1873–1934) are similarly free of undue moral overtones. Indeed, these works are probably less political and moral than those of Comte, Marx, and Durkheim. Thus, it is not difficult to separate Spencerian sociology from his moral philosophy.

2. This was not a sociological work, but a statement of Spencer's general law of evolution, which he felt applied to all realms of the universe— the physical, psychological, organic, and superorganic.

3. I am using the terms used by Talcott Parsons (1968, 1971), since he apparently developed these ideas from his reading of Spencer.

4. Spencer had originally discussed these in *The Study of Sociology* (1873) but elaborated upon them in the opening passages of *Principles of Sociology* (1874–1896, 3–75).

5. *Industrial* was used by Spencer as a synonym for what he also called "operative functions" or internal system processes. *Militaristic* was used instead of *regulative functions* or the power to control external and internal system processes. See *Principles of Sociology* (1876, 458–91). Dates vary for references to *Principles of Sociology*, since various chapters came out in installments to subscribers to Spencer's "Synthetic Philosophy." We have kept to the original publication dates, but page numbers correspond to pagination in the bound volumes. See note 1.

6. Spencer hoped that modern societies would be industrial—that is, concerned with internal productivity—but his observations on the colonialism and warfare of his time often mitigated his hopes. These observations come late in *Principles* (1896) when he describes economic institutions. But if one reads his early, more analytical statements, it is clear that Spencer saw an inherent dialectic between centralization and decentralization in social systems.

7. *Regulatory* = the control of relations between a system and its environment, as well as coordination and control of international system processes. *Operative* = internal system process involved in the "operation" of a system. *Distributive* = the processes by which materials and information move among system units, and between a system and its environment. See *Principles of Sociology* (1876, 498–548).

8. Spencer was quick to point out that society is not an organism, but that there are some similar principles of organization in organic and superorganic systems. As he emphasized in *Principles of Sociology* (1875, 448): "Between society and anything else, the only conceivable resemblance must be due to *parallelism of principle in the arrangement of components*" (emphasis in original). The original analytical distinctions among evolutionary stages appear in volume 1 of *Principles of Sociology* (1876, 549–55). Later volumes fill in the descriptive details.

9. "Simple" stages were divided into those with and those without political leaders, or "heads."

10. In these descriptive statements on various institutions, Spencer was clearly drawing from his accumulating descriptions in his *Descriptive Sociology* (1873–1934), which he had commissioned others to develop under his editorial guidance.

11. It should be emphasized that the present author is, like many others in contemporary sociology, somewhat repelled by Spencer's politics. But his politics should be considered separately from his purely sociological analysis.

12. All of these principles appeared in the first parts of volume 1 of *Principles of Sociology* (1874–1876, 3–75, 447–549). Thus, by 1876, sociology had some of its fundamental laws of human organization.

13. This insight was originally achieved in *Principles of Biology* (1864–1867) and even earlier in several essays.

14. See part 2 of volume 1 (Spencer 1874–1896).

15. Spencer's failure to recognize that cultural ideas are also a major integrative force led Durkheim, and others of the French tradition, to reject Spencerian sociology.

16. Information does not necessarily exceed material resources; only the ratio between them decreases.

17. This insight he took from his *Principles of Biology* (1864–1867) as well as *First Principles* (1862). It also appears in the opening pages of *Principles of Sociology* (1874, 3–38).

18. Obviously, Spencer failed to see the importance of cultural symbols as an integrating force, but even as Durkheim attacked Spencer for this oversight, he borrowed most of Spencer's structural principles. For relevant commentary, see Jones 1974 and Perrin 1975. See also: Turner and Beeghley 1981, Turner and Maryanski 1979, and J. Turner 1978.

4 Spencer's Human Relations Area Files

Talcott Parsons's (1937) opening phrase in *The Structure of Social Action*—"Who now reads Spencer?"—is as true today as it was in 1937. More than ever, Herbert Spencer is portrayed in negative terms, and as a consequence, his corpus of work now receives scant attention. This failure to appreciate Spencer in an age when sociologists worship at the tombstones of Marx, Weber, and Durkheim is an enormous tragedy. For the history of theoretical accumulation in sociology would, we believe, be dramatically improved if Spencer's theoretical ideas had been given the same careful and sympathetic reading as those of other canonized figures of sociology's past. We have made this argument previously (Turner 1985b; Turner and Maryanski 1979); and it need not be elaborated again. Instead, we want to concentrate on Spencer's more empirical work and argue that if his development of an earlier type of Human Relations Area Files had been adopted by sociologists and anthropologists in the decades between 1875 and 1935, both sociology and anthropology would have had broader empirical bases for developing generalizations about human social organization. The obscurity of Spencer's empirical work is so great that the creator of the modern Human Relations Area Files, George Murdock (1965), was once moved to observe:

> This work, so little known among sociologists that the author has encountered few who have even heard of it, inaugurated a commendable effort to organize and classify systematically the cultural data on all the peoples of the world for the advancement of cross-cultural research, and thus clearly foreshadowed the development of the present Human Relations Area Files.

This chapter was coauthored with Alexandra Maryanski and appeared under the title "Sociology's Lost Human Relations Area Files" in *Sociological Perspectives* 31 (1988): 19–34.

To "foreshadow," as Murdock put the matter in the above quote, is very different from "influence." There is not a great deal of difference between Murdock's and Spencer's respective HRAF files, except for the quantitative aspects of the former that have come with modern statistical analyses and the computer. Yet Spencer's HRAF files did not exert great influence on sociology and anthropology, since aside from Murdock, Spencer's *Descriptive Sociology* is hardly ever mentioned. For example, Murdock's teacher at Yale, Albert G. Keller (1915), drops only an occasional footnote to Spencer's *Descriptive Sociology*, as does Keller's older benefactor, William Graham Sumner. Yet Sumner's (1906) famous *Folkways*, which is filled with ethnographic data, does not even reference *Descriptive Sociology* in its bibliography; and even Sumner and Keller's (1927) four-volume text *The Science of Society*, which contains even more ethnographic and historical data, makes infrequent reference to Spencer's empirical work.

Thus in the decades between 1875 and 1935, when Murdock (1940) began to visualize what Spencer recognized fifty years earlier, scholars sympathetic to Spencer hardly noticed his *Descriptive Sociology*. Yet these five decades were crucial years in the history of sociology and anthropology, for it was during this period that they both became firmly established as academic disciplines. But they did so without drawing great inspiration from one of the most important works of the period. We think that the implications of this failure to draw upon Spencer's empirical work are worth pursuing.

Typically, the history of a discipline traces those great events that still influence the present pattern of a scholarly activity. In contrast, we propose to examine a great event that appears to have had very little influence on contemporary sociology and anthropology, and then ask the question: What if Spencer's empirical analysis had been more influential? Such an exercise can tell us a great deal about the history of sociology and is therefore worthy of our attention. Let us begin by summarizing the nature of these files, and then we will explore the implications of their obscurity.

SPENCER'S APPROACH TO ORGANIZING DATA

Spencer developed three different approaches for summarizing the vast amounts of ethnographic and historical data that appear in his works. Indeed, his *Principles of Sociology* (1874–1896) contains over two thousand pages of ethnographic data illustrating his more analytical points. Of the three approaches to arraying ethnographic data, one is most often identified with Spencer: evolutionism. For although Spencer used the term *evolution* to denote the more generic process of structuring,[1] he also employed the term in the traditional sense: stages of societal development from simple to complex. The data on different types of societies are thus arrayed in a way to describe the stages of societal evolution from simple hunters and gatherers through horticulture and agriculture to modern industrial systems. Another approach employed by Spencer for summarizing the data on different societies revolves around the analysis of cyclical phases of societies from centralized ("militant") to decentralized ("industrial" in the general sense of relatively unregulated operation of sustaining processes) profiles. This second approach is often viewed as yet another unilateral evolutionary statement, but in fact it is just the opposite: It analyzes the cycles of the structuring process of human organization. Indeed, it anticipates by several decades Vilfredo Pareto's analysis of "lions" and "foxes" in the "circulation of elites." The third approach, which has been almost completely lost, involves developing broad analytical categories for describing processes in all types of societies. In fact, as the above quote from Murdock illustrates, most contemporary sociologists and anthropologists do not even know that it exists.

Labeled "descriptive sociology," this large body of empirical work develops some general categories for cataloguing historical (if available) and ethnographic data on diverse types of societies, from the simplest to the most complex. Spencer felt that if he could create a general and comprehensive set of categories, these would guide researchers in the recording of data. Moreover, because the categories are the same for all types of societies, it becomes possible to compare societies in a systematic way. As Spencer noted in the preface to one of the volumes of *Descriptive Sociology:*[2]

> In preparation for *The Principles of Sociology,* requiring as bases of induction large accumulations of data, fitly arranged for comparison, I, in October, 1867, commenced by proxy, the collection and organization of facts presented by societies of different types, past and present: being fortunate enough to secure the services of gentlemen competent to carry on the process in the way I wished. Though this classified compilation of materials was entered upon solely to facilitate my own work; yet, after having brought the mode of classification to a satisfactory form, and after having had some of the Tables fill up [with data], I decided to have the undertaking executed with a view to publication: the facts collected and arranged for easy reference and convenient study of their relations, being so presented, apart from hypothesis, as to aid all students of Social Science in testing such conclusions as they have drawn and in drawing others.

In all, there were fifteen volumes of *Descriptive Sociology.*[3] The volumes employed a category scheme developed by Spencer, but the actual compilation of the data from diverse sources was done by others. Upon his death, Spencer left sufficient money in his will to have additional volumes completed. And so, several volumes of *Descriptive Sociology* were completed after Spencer's death. This was possible not only because of Spencer's bequest of money to the project, but, more important, because later editors had very clear guidance from Spencer's system of categories.

The volumes of *Descriptive Sociology* are oversized productions, and almost all open with a tabular presentation of the data. The tabular presentation runs across at least two facing pages (very large pages indeed, as the volumes are so oversized). If several societies are to be examined in one volume, then this open-faced presentation is repeated for each society. There are only slight variations in these tabular presentations of data, which, of course, should be expected, as the whole idea of *Descriptive Sociology* is to sustain a common system of classification. The major variation is the addition of subcategories to include details of more complex societies, but these subcategories always fall under more general categories used to portray simpler societies. The basic rationale for this procedure is best summarized by Spencer (Spencer and Duncan 1874, preface).

> In further explanation I may say that the classified compilations and digests of materials to be thus brought together under the title *Descriptive Sociology* are intended to supply the student of Social Science with data standing towards his conclusions in a relation like that in which accounts of the structures and functions of different types of animals stand to the conclusions of the Biologist. Until these had been done, such systematic descriptions of different kinds of organisms, as made it possible to compare the connexions [*sic*], and forms, and actions, and modes of origin, of their parts, the Science of Life could make no progress. And in like manner, before there can be reached in Sociology generalizations having a certainty making them worthy to be called scientific, there must be definite accounts of the institutions and actions of societies of various types, and in various stages of evolution, so arranged as to furnish the means of readily ascertaining what social phenomena are habitually associated.

Thus for Spencer, *Descriptive Sociology* is to be much like the Linnean classification system in biology. By classifying the features of different societies under common categories, it will be possible to see what "social phenomena are habitually associated" and thereby make generalizations about the operation of societies. Elsewhere, the tabular format, and hence the categories of *Descriptive Sociology* (Turner 1985b, 98–99), have been extracted. In figure 4.1, we have arrayed the categories in a way that captures their underlying structure. But it must be remembered that each of these categories represents a column in the tables of *Descriptive Sociology* where brief summaries of data are recorded. In a sense, the categories and their tabular organization represent a checklist, forcing the compiler to describe structures and processes in a society for each heading and subheading. Let us now review some key features of the tabular form and the underlying category system.

At the top of each oversized table is a short summary of the inorganic environment (geology, topography, and climate), the organic environment (vegetable and animal life), the sociological environment (history, contacts with distant societies, past societies from which current ones evolved, present neighbors), physical characteristics of the population (skin color, height, weight, hair, teeth, strength, and so on), emotional characteristics of the people (e.g., gregariousness, aggressiveness, shyness), and intellectual characteristics of the population (e.g., curiosity, memory, ingenuity). The descriptions of the inorganic, organic, and sociological environments are typically accurate and useful, but the summaries of the characteristics of the people suffer from evaluative overtones. Such is especially the case when describing non–Western populations. Part of the problem is that these descriptions draw upon accounts of travelers, traders, colonial administrators, ministers, and other nonprofessionals, most of whom were convinced of their own superiority and who tended to look at the physical and behavioral patterns of non–Western peoples through very biased eyeglasses. But the descriptions of the physical, organic, and sociological environments are objective.

The main body of the chart is the columns filled with brief summaries. The two major categories are "structural" and "functional," which do not seem to make much sense and so we have eliminated them in figure 4.1. They do not follow from the discussion of function in *Principles of Sociology* (1874–1896) and can, we feel, be ignored. The next subdivision is between "regulative" and "operative"; and here, this division follows from his more theoretical concerns in *Principles of Sociology* (Spencer 1874–1896).

Under "regulative," six categories (at least in volume 3 on primitive societies) are presented: political, ecclesiastical, ceremonial, sentiments, and ideas. Spencer saw ceremony (rituals, titles, badges, forms of address, and the like) as essentially regulatory; they *control* interaction and social relations. Also, unlike Durkheim's (1893) distorted portrayal of him, Spencer was very much concerned with how symbols regulate action. Note that the description of sentiments, ideas, and language all appear under "regulative." Thus symbols as they manifest themselves in different forms—sentiments, ideas, and language—are very prominent in Spencer's system of regulatory processes. The more obvious regulatory structures—polity, law, religion—are also prominent, but no more so than these symbolic processes.

Under the operative processes, Spencer subdivides this general category into processes and products. In volume number 3 of the series, the

FIGURE 4.1. The Category System of Spencer's Descriptive Sociology

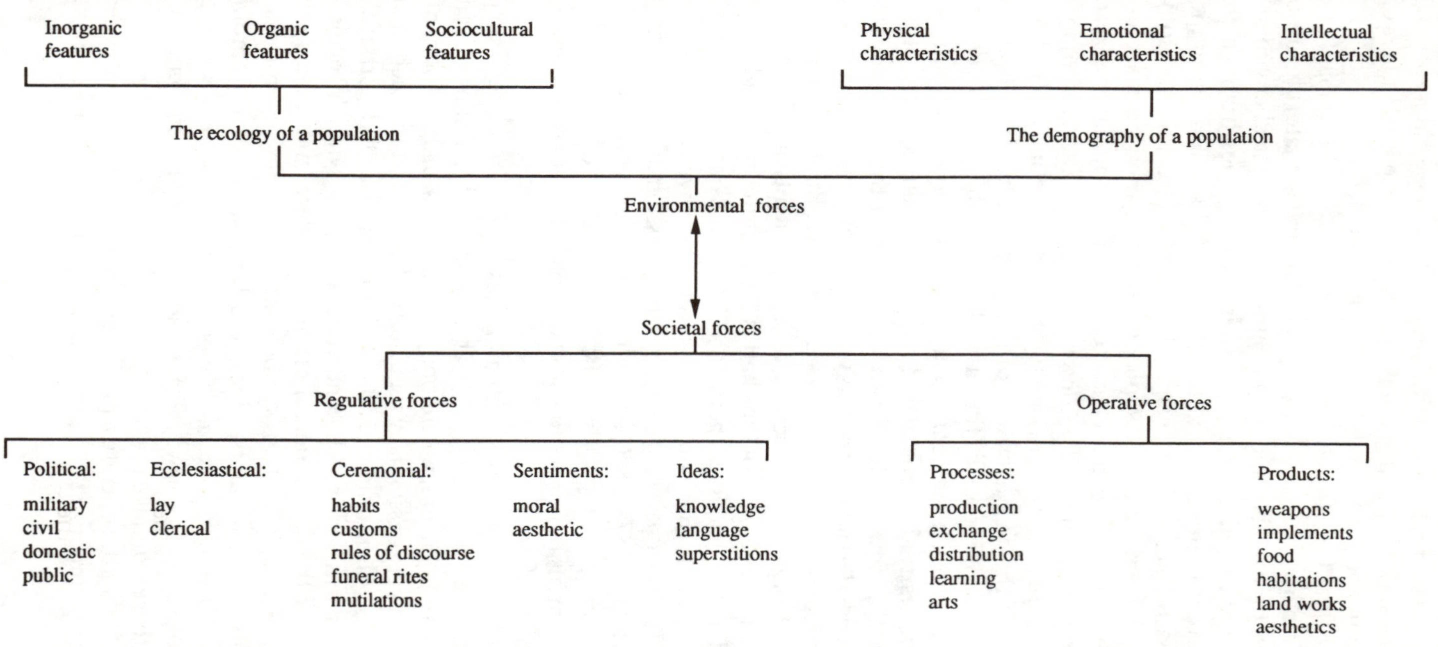

Inorganic features

Organic features

Sociocultural features

The ecology of a population

Physical characteristics

Emotional characteristics

Intellectual characteristics

The demography of a population

Environmental forces ⟷ Societal forces

Regulative forces

Operative forces

Political:
military
civil
domestic
public

Ecclesiastical:
lay
clerical

Ceremonial:
habits
customs
rules of discourse
funeral rites
mutilations

Sentiments:
moral
aesthetic

Ideas:
knowledge
language
superstitions

Processes:
production
exchange
distribution
learning
arts

Products:
weapons
implements
food
habitations
land works
aesthetics

process of distribution is not given the prominence it is given in other volumes on more complex societies (as the distribution systems of the simple societies in number 3 are not elaborate). Similarly, other categories (exchange and production) are not extensive, because of the types of societies explored in volume 3. These and other categories do not change, however. Rather, the columns become wider and filled with more material as more complex societies are examined, but the place of the categories in the classificatory scheme remains much the same in all fifteen volumes.

In these tabular presentations, there is an explicit strategy for analysis. In Spencer's words (Spencer and Duncan 1874, preface):

> Respecting the tabulation, devised for the purpose of exhibiting social phenomena in a convenient way, let me add that the primary aim has been so to present them that their relations of simultaneity and succession may be seen at one view. As used for delineating uncivilized societies, concerning which we have no records, the tabular form serves only to display the various social traits as they are found to co-exist. But as used for delineating societies having known histories, the tabular form is so employed as to exhibit not only the connexions [sic] of phenomena existing at the same time, but also the connexions [sic] of the phenomena that succeed one another. By reading horizontally across a Table at any period, there may be gained a knowledge of the traits of all orders displayed by a society at that period; while by reading down each column, there may be gained a knowledge of the modifications which each trait underwent during successive periods.

We can illustrate this strategy further by comparing volume 1 on "The English" (see note 3) and volume 3 on "Types of Lowest Races." In volume 3, there are successive tables on seventeen different societies: Fuegians, Andamans, Veddahs, Australians (aborigines), Tasmanians, New Caledonians, New Guineans, Fijians, Sandwich Islanders, Tahitians, Tongans, Samoans, New Zealanders (aborigines), Dyaks, Javans, Sumatrans, and Malagasy. By reading across each table, one can see the structural affinities within a society. By reading down a column for all or some portion of the seventeen societies of a given area, one can compare societies for that trait. Moreover, because the societies are grouped by geographical areas (at least approximately), successive tables represent societies in proximate geographical regions. As these societies do not have a written history, only one table per society is possible. In contrast, the volume on "The English" has successive tables for just one society, because it is possible to trace its history. Thus volume 1 reveals successive tables for prominent historical epochs of one society. And so, reading across the table reveals the affinities of traits for a given historical period, whereas reading down any column offers an overview of the changes in that trait over time.

There were, of course, other schemes for ordering and arraying data, particularly among anthropologists, but none comes even close to revealing the analytical detail or empirical richness of Spencer's *Descriptive Sociology*. This system of categories was fully developed by 1873, some thirty years before Durkheim's (1912) efforts to use ethnographic data and, as we mentioned earlier, well over fifty years before George P. Murdock's Human Relations Area Files began to take form. There are a number of important historical consequences resulting from this failure to see the utility of Spencer's categories, for, in our view, ignorance of Spencer altered the history of sociology as a discipline.

SOME HISTORICAL IMPLICATIONS

Cross-Cultural Analysis and Theorizing

What if Spencer's categories, as delineated in figure 4.1, had been used as a "checklist" by early ethnographers going into the field to collect data? If this had been done, we think that we would have had far superior ethnographies in the decades between 1880 and 1940, and perhaps even today. Ethnographic analysis is most useful for comparative purposes when the data are recorded with a common set of categories. Spencer provided, we think, a very comprehensive set; and if these had been used as the guidelines for recording data, social science would have a richer and more comparable base of data on those populations whose traditional social systems were rapidly disappearing under the impact of colonialism in the late nineteenth and early twentieth centuries. Instead of having to impose the categories retrospectively, as Murdock and his associates (1967) have had to do, cross-cultural analysis could have been much more powerful, since the primary data would have been collected using a common conceptual yardstick.

We believe that much of the relativism and historicism that now pervades sociological and anthropological analysis would be less appealing *if* there had been a better data set on preindustrial systems. For it is inevitable that "every culture will be unique" when the data are presented in an idiosyncratic manner by each ethnographer. Moreover, with a superior cross-cultural data base, sociologists and anthropologists would be less antagonistic to general theorizing that seeks to uncover the common and generic properties of human social organization.

The Partitioning of Sociology and Anthropology

Since Spencer was a sociologist and the first great anthropologists, such as A. R. Radcliffe-Brown (1948) and Bronislaw Malinowski (1944), called their work "sociology," the use of a sociologist's classification scheme for recording data would have prevented the artificial partitioning of sociology and anthropology. Indeed, in the early decades of this century there was considerable overlap of these fields. But as ethnographers began to collect data without the benefit of sociological concepts and categories, or *any* common categories for that matter, the fields began to split apart, something that Spencer, Durkheim, Sumner, Keller, Giddings, Ward, and others who dominated sociology at the turn of the century and the adjacent decades would have found regrettable.

The Rejection of Evolutionary Theory

As a consequence of this split between sociologists studying modern societies and anthropologists studying traditional societies, modern evolutionary thinking is often rather naive, especially its Marxist versions. With some obvious exceptions (e.g., Lenski 1966; Lenski and Lenski 1979), social theorists have only a poor grasp of ethnographies; and as a result, when they talk about historical change, their portrayals of traditional systems, especially preagrarian systems, are rather romanticized and inaccurate. And, as a consequence, since the "ethnographic past" is so clearly distorted, such portrayals make evolutionary arguments, and general theoretical arguments as well, highly suspect.

These problems of misportraying traditional systems were clearly evident at the turn of the century, just as Spencer was descending into obscurity, and, with him, his *Descriptive Sociology.* Freud (1913) and Durkheim (1912), for example, both used Baldwin (not Herbert) Spencer's and F. J. Gillian's (1899) travel log on the aborigines of central Australia for their respective analyses of "totem and taboo" and "elementary forms of religious life." Basing a theory of "primitive" and "primordial" human organization on one "case study" is obviously hazardous, especially Spencer's and Gillian's accounts that overemphasize ritual and ceremonial activities (since these were the "most interesting" things to record in photographs and notes). Had Herbert Spencer's analytical categories been taken seriously at this time, Freud's and Durkheim's portrayals of "the first societies" would be less obviously flawed. As the split between anthropology and sociology widened with each decade of this century, general sociological theories, whether evolutionary or not, suffered from the same ignorance about the structure and dynamics of "primitive" social systems.

While evolutionary theorizing is problematic, it can still be highly useful, as Lenski (1966) and others have demonstrated. One does not need to build highly problematic assumptions of unilinear sequences into evolutionary theories, but as Spencer's theories and data base were abandoned, the only remaining models of evolutionary/historical thought were heavily impregnated with unilinear assumptions. Although Spencer is often portrayed as a naive theorist who posited a unilinear theory of evolution, this is a horribly distorted portrayal of his ideas (see Turner 1985c). It is Marx and Durkheim, more than Spencer, who are naive, unilinear evolutionists. And even Weber with his "rationalization thesis" cannot escape this charge. When coupled with the work of less distinguished but influential figures of this time, such as William Graham Sumner and Albert Keller (1927), their sociological portrayals of social change were highly suspect and ethnocentric. These portrayals are far inferior to Spencer's, although most contemporary sociologists think the opposite (typically, without having read Spencer in any detail). Thus if Spencer's evolutionism in its full detail (see Turner 1985c) and his accompanying *Descriptive Sociology* (see note 3) had been the model of theorizing long-term change, evolutionary theory would be a less dubious mode of thinking in the contemporary view.

The Overuse of Explanatory Functionalism

Another consequence of failing to use Spencer's scheme is that, as contradictory as this statement may initially seem, it led to an excessive reliance on *explanatory* functionalism. As ethnographers tried to find coherence in their data and to interpret facts, they were drawn to functional explanations: trait *x* functions to meet need *y* in the systemic whole. As we have argued elsewhere (Turner and Maryanski 1979), the demise of early evolutionism and diffusionism, coupled with the fact that most populations studied by anthropologists had no history, led to an explanatory crisis in anthropology: how to explain events in a culture when one could no longer view them as evidence of an evolutionary sequence, when one could no longer see them as diffused products from "cradles of civilization," and when one could not even trace their history? The answer was anthropological functionalism: Trait *x* exists because it meets need *y* in the society. However, if anthropology had used functionalism not as an explanatory device but as a *descriptive* tool, it would have been much better off. That is, using the descriptive categories contained in Spencer's sociology would have "ordered the data" without trying to explain them. Correspondingly, sociology would have avoided its functional dark ages. Indeed, if anthropologists such as Radcliffe-Brown and Malinowski had not kept functionalism alive in the first half of the century, functional theorizing would, we think, have died. As a theoretical explanation, functionalism is seriously and fatally flawed; but as a descriptive tool it has some merit. Let us elaborate on this point (Turner and Maryanski 1979, 128–41).

Functionalism asks a very interesting question: What does a society need to survive? Functionalists then create lists of functional needs or requisites that are not particularly useful in explaining *why* and *how* cultural traits exist in a society. But these functional needs are, we believe, useful assumptions in constructing guidelines for what is important to observe in a society. For Spencer, there are functional needs for regulation and operation in a society; and hence they become the two basic analytical criteria for his classification scheme (see figure 4.1). He then asks: What are the basic ways these needs are met? And his answer is the various subcategories protrayed in figure 4.1. Thus functional reasoning often leads to very useful systems of categories that can serve as a checklist for recording important (as defined by functional criteria) events.

But anthropology and, later, sociology did not use functionalism in this way. Rather, it was used increasingly to explain empirical events, despite the fact that it is much better at providing criteria for distinguishing "more important" from "less important" events. This is what the scheme in Spencer's *Descriptive Sociology* does; and had social scientists been more aware of it, they could have realized functionalism's utility as a descriptive tool and perhaps avoided its pitfalls as an explanatory approach.

CONCLUSIONS

What, then, can we conclude from this kind of hypothetical exercise? Obviously, we have overstated the historical consequences in order to emphasize our points. Nonetheless, had Spencer's *Descriptive Sociology* been more widely adopted as a research tool and as a way to array data for theorizing and cross-cultural comparison, sociology and anthropology would be still more closely allied. Moreover, with a better data base, we would be less suspicious of general theorizing, especially in evolutionary analysis. Concurrently, we would be less inclined to retreat into extreme historicism and relativism. And most important, we might have avoided the pitfalls of explanatory functionalism and used this mode of analysis in more profitable ways as a descriptive tool.

All of this is speculation, of course. Yet it tells us something about the history of our discipline.

NOTES

1. He also used the term *dissolution* to denote processes of destructuring in systems.

2. This quote comes from volume 3, compiled by David Duncan.

3. The list of volumes of *Descriptive Sociology* is as follows: (1) *English* (1873); (2) *Ancient Mexicans, Central Americans, Chibchans, Ancient Pe-*

ruvians (1874); (3) *Types of Lowest Races, Negritto, and Malayo-Polynesian Races* (1874); (4) *African Races* (1875); (5) *Asiatic Races* (1876); (6) *North and South American Races* (1878); (7) *Hebrews and Phoenicians* (1880); (8) *French* (1881); (9) *Chinese* (1910); (10) *Hellenic Greeks* (1928); (13) *Mesopotamia* (1929); (14) *African Races* (1930); and (15) *Ancient Romans* (1934). A revised edition of number 3, edited by D. Duncan and H. Tedder, was published in 1925; a second edition of number 6 appeared in 1885; number 14 is Emil Torday's redoing of volume 4. In addition to these volumes, which are in folio size, two unnumbered works appeared: Ruben Long, *The Sociology of Islam*, 2 vols. (1931–1933), and John Garstang, *The Heritage of Solomon: An Historical Introduction to the Sociology of Ancient Palestine* (1934).

5 Émile Durkheim's Theory of Integration in Differentiated Social Systems

Theory in the "natural sciences" typically consists of a series of abstract principles that state the relations among generic properties of the physical or biological universe. For example, current evolutionary theory accounts for speciation of animals and plants in terms of principles that specify relationships among ecological change, mutations, gene flow, genetic drift, and natural selection. To take another example, the formula $F = ma$ states that basic properties of the universe are related in certain fundamental ways. In contrast, if we look at the social sciences, particularly sociology and anthropology, a much different picture of "theory" emerges. There is enormous controversy over whether "theory in the natural science sense" is even possible. And among those who believe that social theory is possible, discussion revolves around such issues as the work of the "great masters" of the last century (e.g., Coser 1977; Giddens 1971), the "best" metatheoretical view of the world (e.g., Blumer 1969; Parsons 1937, 1951), the various "schools of thought" (e.g., J. Turner 1978; Ritzer 1975; Harris 1968), or the biography and history of ideas (e.g., Lukes 1973; Nisbet 1974). Indeed, a review of theory textbooks in anthropology and sociology reveals a de-emphasis on isolating and articulating abstract principles of social organization. Much more prominent are discussions of early masters, schools of thought, controversies, and metatheoretical positions. And if we look at introductory texts, we can see an almost complete omission of even those incipient theoretical

This chapter originally appeared in the *Pacific Sociological Review* 24 (1981): 379–91.

concerns in favor of discussions of empirical topics, whether these be complex organizations, kinship, or small groups.

Why should there be such a difference between the "natural" and "social" sciences in the extent to which theoretical issues guide intellectual activity? There are, no doubt, many reasons for the difference, but one of the most fundamental is the failure to view sociology's and anthropology's early masters as theorists. Such a statement may seem surprising, if not absurd, in light of the fact that theory and introductory texts are filled with discussions of the great masters, such as Spencer, Tylor, Morgan, Durkheim, Marx, Weber, Mead, and the like. Yet, it can be argued that these discussions are rarely theoretical in the natural science sense of history. Rather, they typically consist of summaries of discursive texts, presentations of isolated concepts, or delineations of schools of thought that were inspired by an early master.

The lack of concern with extracting the abstract theoretical principles of these masters is particularly surprising, since they appear to have provided keen insights into basic properties of the universe. Indeed, these masters are read and reread today because we sense that they unlocked some of the mysteries of the social universe. And yet, we have not extracted the theoretical essence of their arguments and have left discussions of their metaphysics, metatheory, schools of thought, historical context, and the like to the historians of ideas. In this essay, a modest beginning is made to correct this deficiency. The work of Émile Durkheim is examined with a strictly theoretical purpose: to extract his most abstract theoretical principles of social system differentiation, integration, and disintegration. For whatever else Durkheim is—an idealist (Harris 1968, 464), a functionalist (J. Turner 1978, 25–28), a founder of multivariate analysis, an advocate of causal modeling (Giddens 1971, 1972)—he was a social theorist who articulated some basic laws of the social universe. This is why we continually read him. But we have not fully appreciated Durkheim and other early masters *as theorists.* This lack of full appreciation has led otherwise astute commentators such as Merton (1968, 47) to argue that sociology as well as anthropology will have to wait for its Einstein because "it has not found its Kepler—to say nothing of Newton, Laplace, Gibbs, Maxwell, or Plank." In contrast to Merton, we will argue that we have been reading and rereading our "Newton, Laplace, Gibbs, Maxwell, or Plank." Our problem has been our failure to recognize Durkheim and others for what they are.

DURKHEIM'S THEORETICAL PRINCIPLES

Almost all of Durkheim's major insights (1893, 1895, 1897, 1912) are expressed in an evolutionary, causal, and functional mode of reasoning. It is often the mode of reasoning more than the substantive theory that dominates commentaries on Durkheim (e.g., J. Turner 1978; Turner and Maryanski 1979; Harris 1968; Giddens 1971, 1972). My goal here is to ignore Durkheim's mode of reasoning and translate his substantive argument into theoretical principles. Yet Durheim's (1895) methodology presents a number of problems. First, because Durkheim's theoretical statements are often couched in functional terms (structure *x* functions to meet need *y*), it is often difficult to translate his ideas into a theoretical principle of the form *x varies with y*. Second, this problem is compounded by Durkheim's implicit and explicit mingling of moral statements about what should be and what is. Durkheim frequently defined actual events and structures as "abnormal" and "pathological" in terms of his own moral yardstick. As a result, it is sometimes difficult to discern those relationships that Durkheim felt to be basic to the nature of the social world and, therefore, worthy of statement as abstract principles.

Yet, if these limitations are accepted and it is recognized that they force some degree of inference, a series of abstract principles that summarize Durkheim's thought can be developed. Since Durkheim employed an evolutionary framework, his most important theoretical principles always concern the concomitants of increasing social differentiation. His basic theoretical question appears to have been: What are the basic integrative, and disintegrative, properties in differentiating social systems? This question is phrased more abstractly than Durkheim may have intended, but this is exactly what must be done if Durkheim's genius is to be fully appreciated. It is necessary to pull away from subtle nuances, pay less attention to the burning (but now less relevant) issues of his time, abandon much of his functional vocabulary, and seek a consistently high level of abstraction in order to present the full explanatory power of Durkheim's ideas.

With these considerations in mind, the discussion of Durkheim's theoretical principles will be organized into three sections: (1) principles of social system differentiation, (2) principles of system integration, and (3) principles of system disintegration. It is in these principles that Durkheim's theoretical contribution to modern sociological and anthropological theory resides, and the degree to which these principles are considered insightful will determine the extent to which Durkheim's work can still inform contemporary social theory.

Principles of Social System Differentiation

In *The Division of Labor* (1893), Durkheim isolates a series of variables that he believed influence differentiation and specialization in society. His basic idea is that increased "moral density"—that is, contact and interaction among people—escalates competition for resources, forcing social differentiation. Durkheim's conceptualization is often enveloped in Darwinian and Spencerian metaphors, and thus we need to eliminate Durkheim's vocabulary and direct our attention to the postulated relationship among more generic variables. When this is done, the following principle can be abstracted from Durkheim's work on *The Division of Labor*:

I. The level of social differentiation of a population is a positive function of the degree of competition for scarce resources among members of that population, with competition being a positive and additive function of
   A. the size of a population, with this being a positive function of
      1. the net in-migration
      2. the birth rate
   B. the degree of ecological concentration, with this being a positive function of
      1. the extent of constrictive geographical boundaries
      2. the degree of political centralization
      3. the degree of consensus over cultural symbols.

When ideas are expressed in propositions like the one above, there are several initial reactions by critics that should be addressed before the substance of this and additional propositions is explored. One reaction is that Durkheim's argument has been removed from its historical context. Another is that the proposition is "obvious" and "simplistic." Yet another is that it ignores causality and/or functions. In response, we should emphasize several points. First, abstract scientific theory is always ahistorical; it seeks to understand the generic properties of the universe that cut across time and the details of empirical contexts. Second, good theory always simplifies the universe; it seeks to make understandable the complexity of empirical events and to isolate the key underlying dynamics. In many ways, social scientists have been "snobbish" about their universe, assuming that theory must be esoteric and complex. Just the opposite should be the case. And third, theory in the more advanced sciences ignores issues of final (functional) causes and views causality in general as a secondary issue. Theoretical principles such as the formulas $E = mc^2$ and $F = ma$ or the principle of natural selection do not specify causality. Rather, they seek to reveal the affinities among fundamental properties of the universe. They do not preclude a concern with causality when the formula $E = mc^2$ is used as the theoretical insight for constructing a hydrogen bomb or when we are concerned with identifying the sequence of events in the evolution of a species, such as *homo sapiens sapiens*, by natural selection. But causality is a more relevant concern in examining a specific empirical case in a particular context than it is in formulating abstract social theorems.

With these caveats as a background for this analysis, what does the above proposition say about the social universe? Durkheim's principle, which is not vastly different from Herbert Spencer's (1874), argues that social differentiation and competition over resources are fundamentally related. Those generic conditions producing high levels of competition will increase the level of social differentiation (and by implication, vice versa). Durkheim ([1893] 1935, 256–82) specified two such conditions: population size and ecological concentration. In turn, population size is related to migration and birth rates, while ecological concentration is related to the degree to which populations can be kept from dispersion by geographical barriers, political control, or cultural unity (language, values, beliefs, religious dogma, etc.).

While such a proposition is interesting, it is, no doubt, incomplete. For example, Durkheim has little to say about the effects of technology and productivity on social differentiation. Yet, when Durkheim's argument is stated as a principle, we are in a position to highlight such defects and provide necessary supplementation. For this proposition, we would, at minimum, want to articulate propositions on technology, productivity, and natural resources. More interesting than Durkheim's principle of social differentiation, however, are his principles of social integration and disintegration, for it is in the formulation of these principles that Durkheim's enduring contribution to social theory resides.

Principles of System Integration

For Durkheim, social system integration, or, as he phrased it, "social solidarity," can be defined only by reference to what he saw as "abnormal." Anomie, egoism, lack of coordination, and the forced division of labor all represented to Durkheim (1893, 353–410; [1897] 1951, 145–240) instances of poor integration. Thus, "normal" integration represents the converse of these conditions, leading to the following definition of integration as a condition in which (a) individual passions are regulated by shared cultural symbols (1893, 353; 1897, 241); (b) individuals are attached to the social collective through rituals and mutually reinforcing gestures ([1897] 1951, 171; 1912);[1] (c) actions are regulated and coordinated by norms as well as legitimated political structures (1893, 374–95); and (d) inequalities are considered legitimate and believed to correspond to the distribution of talents (1893, 374–89).[2]

For Durkheim, differentiating systems face a dilemma first given forceful expression by Adam Smith and rephrased by Comte (1830,

1851): The compartmentalization of actors into specialized roles also partitions them from one another, driving them apart and decreasing their common sentiments. In *The Division of Labor*, Durkheim (1893, 287) recognized that differentiation is accompanied by the growing "abstractness," "enfeeblement," or "generalization" of the collective conscience. Or, in terms of the specific variables he used to describe the collective conscience, values and beliefs become less "voluminous," less "intense," less "determinate," and less "religious," for only through increasing generality of the collective conscience can actors in specialized and secularized roles hold common values and beliefs. If values and beliefs are too specific, too rigid, too intense, and too sacred, they cannot be relevant to the diversity of actors' secular experiences and orientations in differentiated roles, nor can they allow for the flexibility that comes with the division of labor in society. Hence, moral imperatives become more abstract and general. This basic relationship can be expressed in the following proposition:

II. The degree of specificity in evaluational symbols in a population is an inverse function of the degree of social differentiation among members of that population.

The key to understanding Durkheim's view of integration, then, is the inherent relationship between differentiation of roles and the increasing generality of moral evaluational systems. For as evaluational symbols such as values, beliefs, and religious dogmas become general and abstract, the major basis for integration in comparatively undifferentiated systems is undone. It "must" be replaced by alternative bases of integration, or disintegrative processes will be initiated. Hence, the degree of integration among differentiated roles will be dependent upon the extent to which alternative bases of integration emerge. Durkheim's list of these alternatives can be expressed in the following principle:

III. The level of integration in a differentiated population in which evaluational symbols are generalized is a positive and additive function of
   A. the degree of intra- and intergroup normative regulation and coordination
   B. the degree of subgroup formation around diverse productive activities
   C. the degree of coordination vested in a centralized authority
   D. the degree of organized opposition to centralized authority
   E. the extent to which the unequal distribution of scarce resources corresponds to the unequal distribution of talents
   F. the degree to which sanctions are restitutive

In this proposition, the varying bases of integration in differentiated social systems are delineated. Any one basis of integration, such as the centralization of authority and the use of coercion, would be insufficient. High levels of integration come when weights for all the bases—that is, IIIA–IIIF—are high. And, of course, there will be varying degrees and forms of integration depending upon the relative weights of each of these variables in a particular empirical system. For example, a system high in normative regulation and subgroup formation will reveal a form of integration much different from one in which centralized authority is weighted higher than the other variables.

Most of the variables in proposition 3 come from *The Division of Labor* (1893), with clarification coming from other works (Durkheim 1895, 1897, 1912). Variable IIIA represents a rephrasing of Durkheim's discussion of his "organic and contractual" solidarity (1893, 200–233) and "organic solidarity" (1893, 111) as well as his discussion of "another abnormal form" (1893, 389–95). Variable IIIB represents a rephrasing of the argument in his famous "Preface to the Second Edition" (1904). Variables IIIC and IIIE are extracted from "the forced division of labor (1893, 374–88) as well as from his reading of de Tocqueville (1835) and his Latin thesis, "Montesquieu and Rousseau" (1892). And variable IIIF is taken from the later portions of his discussion of "methods" (1893, 49–70).

Stating Durkheim's argument as a series of conditions that can reveal varying weights in empirical systems obviates much of the moralistic tone of Durkheim's argument. Equally important, we are able to communicate the dialectical argument built into Durkheim's theory. For in those very conditions that promote integration reside disintegrative processes.

Principles of System Disintegration

Disintegration involves the converse of integration (a) when individual passions are unregulated by shared cultural symbols ("anomie"), (b) when individuals are unattached to collectivities through mutually reinforcing gestures and rituals ("egoism"), (c) when action is unregulated and not coordinated by norms and political authority, and (d) when inequalities are not considered legitimate and in correspondence with the distribution of talents ("forced division of labor"). While Durkheim termed disintegrative processes "abnormal" and "pathological," he also saw these processes as inevitable. For example, Durkheim (1897) recognized that certain rates of deviance stemming from anomie and egoism are inevitable in differentiated social systems. Moreover, there is always tension created by the centralization of power and the mobilization of opposition to power. Additionally, normative regulation is never without ambiguity;

subgroup formation is always incomplete; the distribution of rewards never corresponds perfectly to the distribution of talents (at least in people's perceptions, there is never perfect correspondence); and it is only the ratio of restitution to punishment that changes (not the tension that produces punitive sanctions).

Thus, Durkheim's principle of disintegration is simply the converse of that for integration:

IV. The degree of disintegration in a differentiated population in which evaluational symbols are generalized is an inverse and additive function of variables IIIA–IIIF.

Durkheim's most famous discussions of deviance in *Suicide* (1897) and of "abnormal forms" in *The Division of Labor* (1893, 353–410) are but specific instances of these conditions of disintegration. For example, an "anomic" situation arises when differentiation and generalization of evaluation symbols have not been accompanied by normative restraints on individual desires and passions (1893; [1897] 1951, 353; 1897, 241; 1904). "Egoism" is a situation where differentiation and generalization of evaluational symbols have transpired without a corresponding degree of subgroup formation (1897, 241; 1904). The "forced division of labor" (1893, 374–88) is a result of high weights for variable IIIC and low weights of variables IIIE and IIID. Thus, all of the "pathologies," as Durkheim termed them, of differentiated social systems inhere in the basic conditions of integration. While theoretical discussions of Durkheim often highlight these disintegrative processes, it is also important to recognize that they are part of a more general theory of integration.

CONCLUSION

An attempt has been made to extract from Durkheim's works his basic principles of social organization. As is evident, almost all these principles appeared in his first major work, *The Division of Labor in Society* (1893). Subsequent works clarified key concepts, such as "anomie" and "egoism" in *Suicide* (1897) and "occupational groups" in the "Preface to the Second Edition" of *The Division of Labor* (1904). Yet the basic arguments on the fundamental properties of the social world remained unchanged. Even as Durkheim (1912, 1922) sought to understand social psychological processes, his sociologistic position on integration remained unchanged. As the four principles developed here underscore, Durkheim was primarily concerned with the shifting bases of integration in social systems during their differentiation and elaboration. While many of Durkheim's statements represent his hopes and aspirations for the "good society," they nevertheless reveal a number of critical insights into fundamental relations among generic properties of the social universe.

In particular, principles 2 and 3 are the most insightful, since they argue that social differentiation, value generalization, normative specification, centralization of power, subgroup formation, counterpower, resource distribution, and modes of sanctioning are all related in certain fundamental ways. Durkheim's work thus provides sociology and anthropology with some of its basic laws of human organization. Such a conclusion may seem overly optimistic and premature, since social thinkers often believe that the discovery of laws equivalent to those in the "natural sciences" is either impossible or at least a distant prospect. I hope, by phrasing Durkheim's ideas as propositions rather than as causal or functional statements, pessimistic views on the state and prospects of social theory are rendered less compelling. Sociologists and anthropologists have discovered many of the basic laws of the social universe, but we have often failed to recognize them for what they are.

NOTES

1. For a formalization of this aspect of Durkheim's theory, see Collins (1975, 153–54), where the number of actors, the length of their copresence, their mutually reinforcing gestures, their foci of interest, and their ritual activities are all viewed as variables producing solidarity and shared sentiments. These more microdimensions of Durkheim's theory are well stated by Collins and are not presented in this analysis, which examines the more macrodimensions of Durkheim's theory.

2. This last element of the definition is altered somewhat from Durkheim's own formulation, since I have tried to capture what he defined as integration per se. His discussion of "the forced division of labor" was geared to modern societies, and he did not seem to consider as "abnormal" those inequalities in traditional societies that are not based upon natural talent. His main argument for why the forced division of labor would not persist was that people would not accept it as legitimate in the modern age. Thus, his meaning appears to be that when inequalities are not legitimated, they come to be "forced" or to appear forced.

6 Durkheim's and Spencer's Principles of Social Organization

THEORY VERSUS HISTORY OF IDEAS

In contrast to Max Weber's silent dialogue with Karl Marx (Bendix 1968), Émile Durkheim carried on a very noisy conversation with Herbert Spencer. Like Marx, Spencer was not to reply formally, and so in reality, both Weber and Durkheim conducted monologues with their primary antagonists; as a result, they could interpret Marx and Spencer as they wished. Such is particularly the case with Durkheim's portrayal of Spencer, an issue over which Jones (1974) and Perrin (1975) have already done battle.

I would like to enter this fray with a somewhat different orientation than the one used by those who have previously commented on the relation between Durkheim and Spencer. Rather than examine the issues historically and contextually, I propose that we look at them theoretically. That is, rather than cite text, verse, and context, I will emphasize abstract concepts and propositions. The reason for this shift in emphasis is that the more enduring theoretical significance of their ideas has been underemphasized in various commentaries. In looking at Spencer's and Durkheim's arguments, I am committed to what Jones (1974, 342) has labeled a "presentist" view that "selects out those ideas of the past which conform to present standards." That such a view is contrary to the goals of a "history of sociology" I freely admit. The goal here is to treat Durkheim and Spencer as theorists, to extract their useful *theoretical* ideas, and to forget those that are less enduring. The only

This chapter originally appeared in *Sociological Perspectives* 27 (1984): 21–32.

scientific reason to continue reading and citing scholars of the past is that they had enduring ideas that still offer insight into the dynamics of a social system. Tracing the history of ideas is, of course, a valuable intellectual activity in itself, but it is not science: therefore, it is not the guiding orientation of this analysis. I wish to remove history and context in order to highlight the general theoretical ideas of these great thinkers.

THE BASIC THEORETICAL QUESTIONS

To view Spencer as the intellectual peer of Durkheim violates our present sense of their respective contributions to social theory (Perrin 1976; Peel 1972). This sense is the result, I think, of looking at Spencer through Durkheim's critical eyes—of accepting the truth of his monologue. For unlike Marx, who had hordes of intellectual defenders, Spencer remained vulnerable to misrepresentation, especially since he had espoused a moral philosophy antagonistic to sociologists (most, however, seem content to put aside and ignore Weber's nationalistic-fascist political preachings). Like most contemporary sociologists (including myself), Durkheim abhorred Spencer's laissez-faire moral philosophy (1850, 1892–1898) because it underemphasizes the importance of cultural symbols and because it overemphasizes Adam Smith's "invisible hand of order." This abhorrence of the moral philosophy kept Durkheim from fully appreciating the convergence and complementarity of their *sociological* works. And it has also kept present-day sociologists from appreciating Spencer in his own right, for Parsons's question in 1937—"Who now reads Spencer?"—is equally appropriate today (Parsons 1966, 1). But I should emphasize that Spencer's moral philosophy is *not* prominent in his scholarly works on biology (1864–1867), psychology (1855), and sociology (1873, 1874–1896). True, he viewed his moral philosophy and his scientific work as part of an overall synthetic philosophy, but in reality, the two are separate. Thus, if we select out—indeed throw out—Spencer's moral philosophy and concentrate on his *Principles of Sociology* (1874–1896), it is possible to view Spencer in a more favorable light. For Spencer and Durkheim were addressing the same set of related theoretical questions, first given expression by Adam Smith (1776): "How is integration to be achieved in a world of growing specialization?"[1] To phrase the issue in more abstract terms: (1) What general conditions cause the degree of social differentiation in a social system to increase; and (2) What processes operate to integrate the units in a differentiating social system? It is in answering these two questions that Durkheim's and Spencer's respective contributions to sociological theory reside and converge.

PRINCIPLES OF DIFFERENTIATION AND INTEGRATION

Spencer's answer to these questions was written close to twenty years before Durkheim's, and even earlier if one goes back to Spencer's *First Principles* (1862) or *Principles of Biology* (1864–1867). Spencer (1874–1896) saw structural differentiation as caused by growth of a population within a delimited ecological area. Such growth created a number of interrelated structural dilemmas, including; (1) the problem of coordination and control of the increased number of potentially competing units in the system, (2) the problem of maintaining levels of productivity to support and sustain the growing social mass, and (3) the problem of developing efficient modes of distribution of the products of production. While it was Spencer who coined the phrase *survival of the fittest* (1850) long before Darwin read his essay on natural selection to the Royal Society, this line of argument is not highly prominent in his sociological works. Rather, Spencer referred back to his 1862 work, *First Principles*, whose underlying dynamics were borrowed from physics—"movement of matter," "the persistence of force," "the retained motion," "segregation effects," and "multiplication of these effects." Thus, for Spencer, "organic" (biological) and "superorganic" (social) differentiation are special cases of differentiation in all systems of the universe. Yet, in Spencer's (1874–1896) early sociological analysis, he consistently drew examples from his earlier works in biology (1864–1867) to illustrate the generality of his laws of system differentiation.[2] In a sense, he was sociology's first "general systems theorist," and indeed, more recent synthetic efforts on "living systems" (Miller 1978) simply rediscover Spencer's principles of differentiation. Table 6.1 summarizes his principles and will be used as a reference point in subsequent discussion.

While the specific mechanisms by which size causes differentiation are somewhat vague, the general idea is that increased system size generates two forces that can cause differentiation. One causal process is that as more units are added to the system, problems of sustaining each unit increase, because problems of extracting resources from the environment and then distributing them to each unit escalate exponentially. If a growing system of units is to persist in a given environment, it must differentiate so that some units gather resources, others convert them into usable substances, and still others distribute them. Ecological concentration intensifies these problems, because it confines a growing mass to a limited resource base.

The other causal process underlying the relation between size and differentiation is the one Durkheim (1893) borrowed two decades later.[3] Increases in size, especially in a delimited space (whether for ecological,

TABLE 6.1

A Comparison of Durkheim's and Spencer's Principles of Social Differentiation

Durkheim's Principles

I. The degree of structural differentiation among a population is a positive function of the rates of social interaction and competition among the members of that population, with rates of interaction and competition being a positive and additive function of

   A. The size of the population
   B. the degree of ecological concentration of the population
   C. the number and diversity of communications channels
   D. the scope, scale, and variety of transportation facilities

   E. the degree to which a population is politically and culturally unified

The Process of Structural Differentiation

II. The degree of value generalization among members of a population is a positive function of the degree of structural differentiation of that population.

III. The greater the degree of structural differentiation and value generalization in a population, the more likely is

   A. the elaboration and centralization of political authority

Spencer's Principles

I. The degree of structural differentiation among a population is a positive and additive function of

   A. The size of the population
   B. the degree of ecological concentration of the population

   C. the rate of growth of the population
   D. the degree to which previous structural differentiation has been integrated

II. The greater the degree of structural differentiation in a population, the more likely is differentiation to involve

   A. the elaboration and centralization of political authority (regulatory structures), with centralization being a positive function of the degree of (1) external threat to a population, (2) dissimilarity of subgroups in a population, (3) the volume of internal transactions, and (4) the level of productivity

B. the elaboration of nonregulatory productive units (operative processes), with the level of productivity being a positive function of (1) the availability of resources, (2) the level of technology, (3) the degree of IIC

C. the elaboration of separate distributive units as the volume of transactions, with the degree of distribution elaboration being a positive function of IIB and IIA

III. The degree of support for increased elaboration of political authority is a positive function of the level and duration of unrestricted elaboration of productive IIB and distributive IIC processes.

IV. The degree of resistance to political authority is a positive function of the level and duration of political restrictions on productivity IIB and distributive processes IIC.

B. the formation of subgroups that organize members who engage in similar productive activities

C. the development of normative rules for regulatory relations within and among members of that population

D. the elaboration of organized counterauthority

E. the development of distributive mechanisms in order that the unequal distribution of rewards among members of a population corresponds to the unequal distribution of talents

F. the increase in the proportion of restitutive over punitive sanctions

IV. The degree of "anomie" among members of a population is an inverse function of IIIC above.

V. The degree of "egoism" among members of a population is an inverse function of IIIB above.

VI. The degree of inadequate coordination among members of a population is an inverse function of IIIA and IIIC above.

VII. The degree of political instability among members of a population is a positive function of IIIC and an inverse function of IIID and IIIF above.

VIII. The degree of forced division of labor is an inverse function of IIIE above.

political, or cultural reasons, with ecological given the most emphasis), increase competition for resources, which, if left unfettered, would destroy the system. Hence, units differentiate their activities in order to reduce their levels of competition and increase their levels of cooperation in securing, producing, and distributing resources.

Once integration of the diversity in system units has occurred (in accordance with processes delineated in proposition II, and subpropositions IIA, IIB, IIC), then further system growth and differentiation can occur (subproposition ID). But without the increased skeletal structure provided by integration among differentiated units, the lack of integration serves as a source of resistance to further growth. Without some capacity to sustain units, to coordinate their activities, and to mitigate conflict, the system cannot take on more mass. If it does, it will begin to dissolve.

Thus, more recent work in human ecology (Hawley 1950; Hannan and Freeman 1977), organizational growth (Blau 1970; James and Finner 1975; Hendershot and James 1972), social interaction (Mayhew and Levinger 1976), governmental growth (Stephan 1971), and the growth of government in nation-states (Nolan 1979) all invoke these ideas from Spencer's analysis of size and system growth. Thus, many authors have unwittingly been explicating Spencerian sociology.

This fact becomes even more evident when Spencer's analysis of the process of differentiation is added to his causal analysis. In essence, Spencer asks: What is the pattern and form of structural differentiation as efforts to resolve dilemmas of competition, coordination, productivity, and distribution are made? Spencer's answer is presented in proposition II in table 6.1 and revolves around the elaboration of regulatory centers (political authority), the expansion of operative processes (processes of gathering and producing as they influence the internal organization of the system), and the eventual development of distributive processes as the volume of movement, communication, and exchange of goods and services escalates. These processes, Spencer argued, are self-reinforcing: Expanded production requires further elaboration of power and distributive processes to coordinate units and distribute resources; conversely, the extension of political regulation facilitates increases in production and distribution by decreasing the negative effects of competition and poor coordination among system units.

There is also a set of corrective dialectical processes among these forces, as Spencer's often-quoted but frequently misunderstood distinction between "militant" and "industrial" systems underscores (Turner and Beeghley 1981, 108). The consolidation of power (militant systems) begins to restrict operative and distributive processes, creating economic stagnation, restrictions on innovation, and public resentment of regulation. This resistance to further political control causes a decentralization of power and the deregulation of operative and distributive processes (industrial systems). But over time, decentralized systems create problems of coordination and control that escalate to a point where productive and distributive processes are so poorly coordinated that centralization of power becomes essential to sustain the system (hence, transformation back to a militant profile). As Spencer (1874-1896, 569-80) emphasized in a way that Pareto (1901) was later to argue, social systems cycle between militant and industrial profiles as the inherent problems of either a centralized or a decentralized political control successively escalate.

As table 6.1 underscores, Durkheim's causal analysis of differentiation is essentially the same as Spencer's, except that it emphasizes only one causal process: Competition among units increases the selection pressures for differentiation. It adds nothing new to Spencer's model. Where Durkheim surpasses Spencer's analysis is in the recognition that cultural forces are a central dynamic in the process of differentiation.

With structural differentiation—whatever its causes—comes an "enfeeblement" or "generality" of the collective conscience (Durkheim 1893, 170-73), or what Parsons (1966) later called "value generalization." This relationship between differentiation and value generalization is the central dynamic of Durkheimian sociology (see proposition III in table 6.1). Differentiating systems inevitably produce generalized values and, in the process, weaken the major integrating force (the collective conscience) in less differentiated systems (mechanical solidarity). For Durkheim the central moral question, and the one that represents his enduring theoretical contribution, is about the alternative modes of integration that develop in differentiated systems with generalized values. In *The Division of Labor* (1893), as supplemented by the 1904 "Preface to the Second Edition" and as clarified by his more precise definitions of terms in *Suicide* (1897),[4] there are a number of alternative integrating processes (propositions IIIA to IIIF in table 6.1): (1) the development of normative agreements that specify more concretely the way social units are to act and interact, (2) the formation of (occupational) subgroups around key productive activities that bind individuals to a common activity and that allow them to form a common perspective,[5] (3) the development of decentralized authority, which is mitigated by counterauthority associated with productive subgroups, (4) the development of distributive systems in which the unequal allocation of resources corresponds to the unequal distribution of talents, and (5) the codification of formal rules with restitutive, as opposed to punitive, sanctions. For Durkheim, the more conditions 1 through 5 are met, the more integrated a social system is. Conversely, specific pathologies follow from the failure of these conditions to be realized. For example, if condition 1 is not met, anomie ensues; if 2 is not developed, egoism results; if 1, 3, and 5 are not realized, lack of coordination increases (what Durkheim [1893, 389] termed "another abnormal form"); and if 3 and 4 are not in place, a forced division of labor results (see propositions IV to VII in table 6.1; see also Turner 1981).

In a wide variety of contexts, the ideas listed in propositions II through VIII in table 6.1 have been incorporated into present-day sociological thought. The major problem is that Durkheim's moral convictions forced him to see the processes delineated in subpropositions IIIA through IIIF in table 6.1 as normal, and their failure to be implemented as pathological. The theoretical as opposed to moral question thus becomes: Under what generic conditions is the degree or level of these processes increased or decreased?

In sum, the propositions in table 6.1 facilitate observation of the points at which Durkheim and Spencer converge and diverge in their thinking. Durkheim's and Spencer's models of the causes of differentiation are virtually the same; and thus, it is in the process of structural differentiation, once initiated, that we can observe how their principles complement each other. Spencer's principles on the process of structural differentiation stress the dynamics of power, production, and distribution. Durkheim's principles emphasize the problem of moral constraint and the dynamics that follow from highly generalized values in structurally differentiating systems.[6] In this short note and in the format of table 6.1, I do not attempt to synthesize Spencer and Durkheim; rather, the goal is to extract their basic theoretical principles. This extraction is not intended to represent a synthesis of two historical figures, but is instead an invitation to use the ideas in table 6.1 as a means for developing principles of two fundamental processes in human social systems: differentiation and integration. I suspect that in these propositions in table 6.1 lie some of sociology's laws of the social universe. And it is this fact that pulls us back to Durkheim's work and, one hopes, to Spencer's.

NOTES

1. In the opening page of *The Division of Labor*, Durkheim gives A. Smith (1776) and J.S. Mill credit for formulating the problem of solidarity in differentiating social systems; yet Durkheim's attacks are saved primarily for Spencer, who was viewed by Durkheim as the sociological descendant of A. Smith and J.S. Mill.

2. The underlying mechanism for both "organic" (biological) and "superorganic" (social structural) differentiation is a first principle that analogizes to physics; that is, social differentiation is a process of "integration of matter and concomitant dissipation of motion; during which the matter passes from an indefinite incoherent homogeneity to a definite coher-

ent heterogeneity; and during which the retained motion undergoes a parallel transformation" (Spencer 1862, 343).

3. Durkheim claimed that he was taking this idea from Darwin, but it is really Spencer's. In fact, Spencer coined the phrase "survival of the fittest" long before Darwin, who gives Spencer special credit in the introduction to *On the Origin of Species*.

4. Durkehim's (1893) early definition of anomie is vague and is only clarified in *Suicide* (1897), when he distinguishes anomie (lack of moral constraint on aspirations) from egoism (lack of involvement in group processes).

5. This idea comes from the preface to the second edition of *The Division of Labor*, published in 1904.

6. This is about as close as Durkheim comes to acknowledging Marxian sociology (compare with all the references to Spencer).

7 Émile Durkheim's Final Theory of Social Organization

Commentary on the work of early theorists is one of the mainstays of contemporary sociological theory. Indeed, the term *metatheory* has been invented to acknowledge this intellectual tendency among current theorists. Yet the term has become a general "gloss" for just about any kind of commentary, and as a result, metatheory rarely produces scientific theory. Instead, it generates philosophical and historical discourse that becomes, in itself, a self-sustaining activity.

In contrast to this tendency to metatheorize on Durkheim (e.g., Jones 1986; Alexander 1982; Lukes 1973), this chapter will examine those portions of Émile Durkheim's work that can potentially produce scientific theory. To this end, I begin with a review of some problems in representing Durkheim's ideas scientifically; then I will construct a complex causal model of those portions of Durkheim's work that are amenable to theorizing; and finally, I will selectively translate the model into a series of abstract propositions that can legitimately be termed "Durkheim's Laws."

PROBLEMS IN ARTICULATING DURKHEIM'S THEORY

Substantive Problems

One substantive problem that immediately surfaces is Durkheim's nonscientific advocacy of a particular type of "moral society" (e.g., Durkheim

This chapter originally appeared under the title "Émile Durkheim's Theory of Social Organization" in *Social Forces* 68 (1990): 1089–1103.

1922). If these ideological commentaries stood by themselves in separate
volumes or passages, they could be easily ignored; but unfortunately,
they are woven into his more scientific efforts to produce explanatory
laws. For example, throughout Durkheim's discussion of social differentiation and integration are more moralistic statements on education, politics, and inequality; and in fact, these statements are so ideologically in-
fused that they cannot be incorporated into more neutral theoretical
statements. Other topics— anomie, egoism, normative regulation,
solidarity-integration, and differentiation—are morally laden, but it is still
relatively easy to extract less evaluative theoretical statements. Thus, in
this exercise, I will be selective and focus only on those processes that can
be separated from Durkheim's moral commitments. This will make some
of the major deficiencies of Durkheim's theory evident, specifically, the
lack of an extensive conceptualization of the dynamics of power, inequality, and conflict.[1]

Another major substantive problem is Durkheim's inattention to
economic or productive processes—a major oversight in light of the evolutionary thrust of his work. Indeed, "production" is rarely viewed by
Durkheim as a variable; instead it remains implicit and is usually presented through proxy variables, such as transportation and communication technologies. Obviously, a theory of social organization that does not
address the means by which actors are sustained contains a major flaw.

A third substantive problem, which has often been commented
upon (e.g., Giddens 1972), is the shift in Durkheim's work from a macrostructural level in the early 1890s (Durkheim [1895] 1938, [1893] 1933)
to an ever more social-psychological and interpersonal emphasis
(Durkheim [1912] 1965, [1897] 1951; Durkheim and Mauss [1903]
1963). Yet, in contrast to some commentators, I see this shift not as a
problem, but as an interesting theoretical challenge: to integrate the mi-
cro and macro levels of the theory in a manner that corresponds to
Durkheim's intent.

Conceptual Problems

In his most formal statement, Durkheim ([1895] 1938, 95) presents a most
problematic view of sociological explanation:

> When the explanation of social phenomena is undertaken, we must seek
> separately the efficient cause which produces it and the function it fulfills.

In arguing in this way, Durkheim ignored one of his intellectual mentors,
Auguste Comte, who insisted that it is "vain" to conduct "research into
what are called *Causes*, whether first or final" (Comte 1830, 5, italics in

original). Yet, while ignoring Comte's advice, Durkheim still sought to adhere to Comte's (1839, 5–6) view that scientific sociology "regards all phenomena as subject to invariable natural *Laws*" and that "our real business is to analyze accurately the circumstances of phenomena, and to connect them by natural relations of succession and resemblance." In Comte's (1830, 6) assessment, the "best illustration of this is in the case of the doctrine of Gravitation"—presumably Newton's famous formulas, $f = ma$, or $F_g = (Gm_1 m_2/r^2)$.

Thus, a major problem in presenting Durkheim's theory is coping with what he means by explanation. The most difficult task is to sort out various types of causal explanations and reconcile them with (1) his articulation of noncausal laws, like Newton's principles of gravitation, which leave causality ambiguous,[2] and (2) his statements about functions or how phenomena operate to meet needs for social integration.

One solution to this problem is to translate all functional statements into laws that simply state, like Newton's equations, a particular pattern of relation among variables—in Newton's case, mass, distance, and gravitational attraction; or mass, acceleration, and force. I will not present mathematical equations, but in principle, Durkheim's laws reveal the logic of equations without the corresponding precision. For example, Durkheim ([1893] 1933, 262) often makes statements like "the division of labor varies in direct ratio with the volume and density of societies." Such statements are not causal arguments, but they are easily translated into equations.

I will also translate Durkheim's ideas into a causal model, in which effects of variables on each other are arrayed. This latter task is particularly difficult, because Durkheim's theory reveals varying and vague conceptions of causality. Without being pulled into the philosophical quagmire that has always surrounded attempts to sort out causes,[3] let me indicate some of the types of causal arguments in Durkheim's work.

Many of Durkheim's functional statements, which might otherwise be viewed as tautologies or illegitimate teleologies, can be converted into less problematic statements by invoking a "social selection" argument, as is done with functional statements in biology.[4] For example, rather than imply, as Durkheim often does, that the "need for integration" directly causes the emergence of the division of labor (an illegitimate teleology),[5] it can be argued that certain kinds of structures and processes are retained because they had (and have) selective advantage over others for promoting social integration. It is necessary to recognize, then, that many of Durkheim's theoretical ideas are couched in such selection terms, because he was borrowing metaphors from Spencer (1874–1896) and Darwin (1859). The problem with selection arguments, however, is that the particular cause(s) of these structures and processes is (or are) obscured in favor of moral pronouncement of what *should* occur—that is, it is "good" to have social solidarity and integration.

Another way in which Durkheim treats causality is statistically, as in *Suicide* ([1897] 1951). Here, emphasis is on isolating "the cause" of an event through the successive elimination (through what he saw as statistical controls) of alternative hypotheses on causes. This type of causal analysis is not prominent in Durkheim's theoretical works, except to the extent that it forced him to clarify some important concepts, such as "anomie" and "egoism," which are very important in his theory but which were vaguely conceptualized in earlier works, such as *The Division of Labor* (Durkheim, [1893] 1933).

Far more important than statistically generated "causes" are those statements in Durkheim's analysis that examine "efficient causes." What these turn out to be are statements on the generic or general conditions that produce a basic class of results. For example, Durkheim asserts that competition among individuals is caused by the concentration of a population in ecological space, thereby increasing material and social "density" (Durkheim [1893] 1933, 266). These are not causal explanations of the type: This particular empirical event, perhaps in conjunction with other specific empirical events, sets off another concrete empirical event, say, the French Revolution. Instead, Durkheim's "efficient causes" are more abstract, stating a general condition that produces another general type of event. This is what makes them theoretical, but it is also what makes them less precise, since it is no longer possible to trace precisely how one empirical condition sets another into motion. Moreover, these statements are often weakened further by evolutionary assumptions that certain sequences of events are inevitable. Here, it is very difficult to discern how one condition produces another condition, which then causes yet another, and so on; rather, a sequence is simply described, and the causal connections in the sequence are left implicit.

Yet another way in which Durkheim addresses causality is through complex statements of multiple causes: A number of events—say A, B, C—are seen as operating simultaneously, and perhaps interactively, to bring about a given outcome—D. For example, transportation and communication technologies, in conjunction with ecological concentration, are believed by Durkheim to increase the number of individuals in interaction (Durkheim [1893] 1933, 260). There is a sense in which these variables simultaneously produce this effect, but the result is a vague causal statement. Moreover, the implied interaction effects among transportation, communication, and concentration make the causal statement even more imprecise; and as a consequence, it becomes difficult to pinpoint causal effects, except in an aggregate way.

An additional type of causal argument in Durkheim's work is what

Stinchcombe (1968) has called "reverse causal chains." In such causal statements, the outcome of causal sequence determines the subsequent flow of this sequence. For example, the emergence of a division of labor out of "competitive struggles" among individuals "feeds back" and decreases the intensity of the struggle and provides a basis for social integration among potentially competitive actors (Durkheim, [1893] 1933, 266–69). These are not true "feedback" arguments, because goals, purpose, and parameters are not conceptualized as components of a self-regulating system. Since Durkheim had such a poorly developed conceptualization of power and political regulation—except for some rather grandiose ideas adapting Tocqueville to Montesquieu (1748)[6]—a true set of cybernetic processes revolving around system goals, output, feedback, and correction cannot be imposed on Durkheim's theory. I will, at times, loosely refer to these reverse causal chains as "feedback," but I do not mean by this term a true system of cybernetic self-regulation.[7]

A final type of causal argument in Durkheim's analysis is by "causal mechanism": A given set of conditions causes a particular process to be activated; in turn, this process operates in ways that make certain outcomes more likely. These kinds of causal arguments are more complex, because the mechanism is often a cluster of processes, that, for the explanation at hand, goes unanalyzed. For example, Durkheim's ([1893] 1933, 266) analysis of ecological concentration, struggle, and competition (as the selective mechanism) and the emergence of the division of labor is an argument in terms of a mechanism. In this case, two events (concentration and specialization) are connected via a rather vague set of processes (competition and struggle) operating as a mechanism.

In sum, these problematic issues present us with points of ambiguity in Durkheim's theory. I have dwelled on them because they help explain why I go about analyzing Durkheim's theory in a certain way. As noted earlier, I will begin by converting Durkheim's ambiguous statements into a complex causal model; then I will translate portions of the causal model into sociological laws that, by their nature, leave the question of causality implicit and, instead, highlight the form of relationships among variables—much as in the case of Newton's laws on attraction, force, and gravity. This movement between complex causal models and lawlike statements is a useful theoretical strategy (Turner 1988, 1987, 1985a), because it allows us to see complex configurations of causal connections and, at the same time, to postulate what forces in the social universe reveal lawful relations to each other, without introducing the complexity and philosophical ambiguities that always plague causal analysis. Moreover, models and laws can serve as correctives for each other: Laws beg for models outlining causal connections informing us why and how variables are related in a particular way (e.g., the formula $F_g = [Gm_1m_2/r^2]$

led to a search for the causal processes or the mechanism accounting for the "pull" or "force" of gravity), whereas complex models need to be translated into simpler statements involving fewer variables if they are to be tested (since complex models cannot be empirically assessed as a whole).

DURKHEIM'S DYNAMIC MODEL OF SOCIAL ORGANIZATION

If we combine the entire corpus of Durkheim's work into a parsimonius model, what would his model look like? Figure 7.1 presents my best efforts along these lines. Figure 7.1 is not a theory per se, but a dynamic modeling scheme. In constructing such a model, the goal is to (1) selectively translate Durkheim's ideas into generic classes of variables and (2) highlight various causal paths that can suggest abstract propositions or laws. This model can be considered "dynamic," because it emphasizes the flow of causal processes through direct, indirect, and feedback (actually, reverse causal) paths. As the signs on each causal path denote, Durkheim posited a certain type of relationship among the variables.

Before proceeding, however, I should emphasize several problems with this model. First, it is too complex to ever be tested empirically; and even if we sought to employ simulations, it might still be too complicated. At best, we can use portions of the model (particular causal paths) for tests and simulations. Second, there are many areas of ambiguity in the nature of the causal connections denoted by the arrows and signs. These are partly the result of Durkheim's portrayal of cause (as outlined earlier), but also the inevitable result of *any* causal analysis that, by its very nature, will be problematic. Third, while Durkheim's concepts are all converted to variables (and are somewhat relabeled), his definitions are often imprecise, forcing a certain amount of guessing as to what the variables denote. Fourth, the model is too endogenous and filled with self-reinforcing causal cycles. For adequate tests of the model to occur, we would need to add exogenous forces as these influence the values for the variables in ways that reduce the closed, mutually reinforcing cycles.

Yet, despite these problems, the model is a useful heuristic. It pieces together Durkheim's major theoretical ideas and delineates the configuration of processes that Durkheim saw as crucial. And it does so without the functional and moralistic overtones that typify much of Durkheim's work.

The most characteristic feature of figure 7.1 is that it outlines an ecological theory, because the instigating causal forces are population size, growth, movement, communication, and concentration. Such is the case at both the macro- and microlevels, since it is the concentration of

FIGURE 7.1. Durkheim's Model of Social Organization

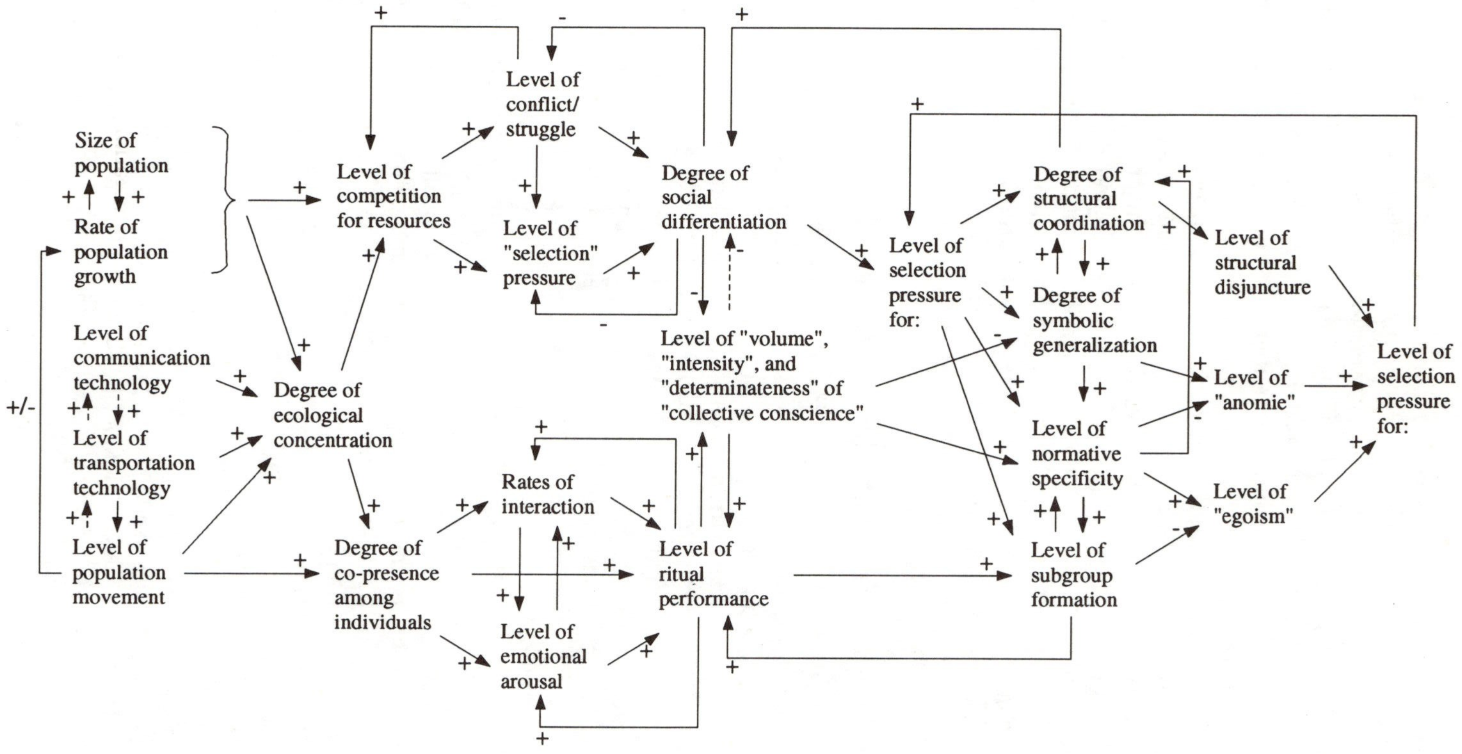

individuals in space that sets into motion both the macrolevel processes of competition, conflict, selection, and differentiation as well as the microlevel processes of copresence, interaction, emotional arousal, and ritual. Thus, ecological variables initiate various macro- and microprocesses, which are delineated, respectively, across the top and bottom portions of the model. But, once initiated, other important causal dynamics become operative, as is illustrated by the causal paths in the middle and right portions of the model. Another feature of the model is that macro- and microlevel processes are "linked" together in a complex set of causal connections among those variables influencing the degree of integration in a social system (the right portions of the model). Hence, it is when Durkheim addresses the question of integration that we can see how his early macrostructural ideas can be reconciled with his later micro-interpersonal focus. With these general features in mind, then, let me begin with macrostructural processes, then move to the micro-interpersonal, and finish with how the two models can be reconciled.

Durkheim's Model of Macrodynamics

The Division of Labor ([1893] 1933) is, of course, the main source of Durkheim's macrostructuralism. Basically, Durkheim views the concentration of a population in ecological space—what he termed "material density"—as crucial to the division of labor, with population size and rate of growth as central "causes" of such density. But ecological conditions exert an independent effect, since the same sized population can be concentrated or dispersed as a result of varying amounts of space and geosocial configurations (natural barriers, cities, etc.). Material density is also "caused" by "the number and rapidity of ways of communication and transportation" (Durkheim [1893] 1933, 259) because these technologies "suppress" and "diminish" the "gaps separating social segments," thereby increasing the concentration or "density" of contact among actors in ecological space. Similarly, migration and mobility, as "caused" by transportation technologies, increase contact and potentially provide a means for further concentration of a population. Since population growth, and its effects on size and ecological concentration, are affected by net migration (immigration less emigration), this variable exerts an indirect causal effect on concentration through its influence on population growth. Thus, Durkheim posits a "multiple cause" argument with some interaction effects for increased ecological concentration, or "material density." Presumably each variable exerts an independent effect, but together the variables exert an even greater effect.

Ultimately, the "division of labor," or what I term "social differentiation," is "caused" by the increased "moral density" that follows from escalated "material density." It is at this point that Durkheim's causal argument gets slippery, because he invokes a selection argument, positing natural selection as a causal mechanism that translates ecological concentration into a division of labor, or increased social differentiation. This occurs because the "struggle for existence is more acute" (Durkheim [1893] 1933, 266), thereby increasing selection pressures. Durkheim implies that concentration increases competition for resources, which, in turn, causes this "struggle for existence," when he notes (ibid.):

> Darwin justly observed that the struggle between two organisms is as active as they are analogous. Having the same needs and pursuing the same objects, they are in rivalry everywhere. As long as they have more resources than they need, they can still live side by side, but if their number increases to such proportions that all appetites can no longer be sufficiently satisfied, war breaks out, and it is as violent as this insufficiency is more marked; that is to say, as the number in the struggle increases.

I have also drawn a causal arrow directly from the population variables to competition for resources to acknowledge what this passage emphasizes: Population size, per se, exerts a direct influence on scarcity of resources. Competition for scarce resources thus causes conflict, under conditions of material density; and it is at this point that Durkheim invokes a selection mechanism, presumably because competition and conflict create selection pressures for specialization. After offering biological examples on speciation, Durkheim asserts that "men submit to the same law" (Durkheim [1893] 1933, 267) and concludes that "it is easy to understand that all condensation of the social mass, especially if it is accompanied by an increase in population, necessarily determines advances in the division of labor" (Durkheim [1893] 1933, 268). Thus "social speciation" or "specialization" occurs as a result of the mutually escalating effects among competition, struggle, and selection pressures (note direct, indirect, and reverse causal arrows among these processes in the model).

Once specialization or social differentiation exists, it exerts a feedback effect (actually a reverse causal effect, unless "directed" by a regulatory agent such as government) on conflict and selection pressures, for while "the division of labor is . . . a result of the struggle for existence . . . thanks to it, opponents are not obliged to fight to a finish, but can exist one beside the other" (Durkheim [1893] 1933, 270). Hence, I have drawn feedback (or reverse causal) arrows back to conflict and selection pressures.

Social differentiation, in turn, changes the nature of integration in social systems. Durkheim's discussion of these changes marks, I think, his most important theoretical contribution to macrostructural analysis (since his argument, thus far, simply repeats Herbert Spencer's position without Spencer's sense of the effects of power and regulation on selection processes). For all its insight, however, this portion of Durkheim's theory has to be disentangled from his moral preachings on the "good society" and his rather extreme functionalism. This disentangling is further complicated by an implicit selection argument that is difficult to portray in a causal model, although I have sought to do so. Nonetheless, the right portion of figure 7.1 represents my best effort to delineate the causal processes as I think Durkheim conceived them.

Social differentiation creates selection pressures for a decrease in the "volume," "intensity," and "determinateness" of the collective conscience (Durkheim [1893] 1933, 152, 167)—that is, collective symbols, such as values and beliefs, become less likely to be shared ("volume"), less powerful and constraining ("intensity"), and less clear ("determinateness"). At the same time, and presumably as a result of this "enfeeblement" of the collective conscience (Durkheim [1893] 1933, 171), there are selective pressures for what Parsons (1966) was later to term "value generalization." I have phrased this process "symbolic generalization," because Durkheim uses the term *abstractness*. Hence, the collective conscience

> changes its nature as societies become more voluminous . . . the common conscience is itself obliged to rise above all local diversities, to dominate more space, and consequently to become more abstract. (Durkheim [1893] 1933, 287)

A phrase like *is itself obliged* is causally vague, to say the least, but I would suggest that Durkheim intends a selection argument here: In the competition among symbol systems, those that are general and resonate across the diverse experiences of differentiated units will be retained, especially since there are additional selection pressures stemming from disintegrative tendencies in social systems (see the last variable on the right of figure 7.1).

Yet, as we will see in examining Durkheim's microlevel theory, there are microprocesses working to promote a high "volume," "intensity," and "determinateness" of the "collective conscience" (note arrows from "ritual performance" to "collective conscience"). These processes, which are based upon microencounters, create pressures for a low level of abstraction in cultural symbols. This tension is "resolved" by the creation of subgroups (what Durkheim [1902] termed "occupational groups") and normative specificity within and between such groups. Even before the new preface to *The Division of Labor*, Durkheim ([1893] 1933, 205) emphasized:

> If society no longer imposes upon everybody, it takes greater care to define and regulate the special relations between different social functions.

And goes on to stress (Durkheim [1893] 1933, 302):

> It is certain that organized societies are not possible without a developed system of rules which predetermine the functions of each organ. In so far as labor is divided, there arises a multitude . . . of moralities and laws.

Such processes are created and sustained by interaction rituals at the microlevel, but such rituals are "selected" because of the potentially disintegrative effects of structural differentiation and symbolic generalization. Thus, selection pressures, as they encourage certain kinds of ritual practices, produce the subsolidarities, and normative linkages between them, that mitigate against differentiation and symbolic generalization.

The last causal effect in the model is what Durkheim termed "another abnormal form." I labeled this process "structural disjuncture," because this is what Durkheim had in mind. The basic idea is that social differentiation creates selection pressures for structural coordination—exchange and interdependence among units, structural overlap, and perhaps structural inclusion (units inside more inclusive units). Failure to produce such patterns of coordination creates disjunctures that escalate selection pressures to recoordinate relations among units. If not, the system "dies" through disintegration.

Similarly, other outcomes of the selection pressures emanating from social differentiation can potentially produce "pathologies." As Durkheim clarified in later works, anomie is the result of symbolic generalization without a corresponding normative specification to regulate passions and desires (Durkheim [1897] 1951, 257–58), whereas egoism is the lack of involvement and participation in group structures (Durkheim [1897] 1951, 1902). Any of these "pathological" or disintegrative outcomes, resulting from the failure of the selection pressure emanating from social differentiation (and the failure to produce rituals at the microlevel), escalate selection pressures for a particular pattern of integration; structural coordination (for structural disjuncture), normative specification (for anomie), and subgroup formation (for egoism).

Durkheim's Model of Microdynamics

By the time that Durkheim had turned to suicide (Durkheim [1897] 1951), he had become concerned with social psychology. Later, when he wrote on religion, he could argue that "the collective force is not entirely outside of us . . . this force must also penetrate us and organize itself within us" (Durkheim [1912] 1965, 209). Whatever the merits of Durkheim's speculations on the origin of religion, his work on this topic contains a powerful theory of microdynamic processes. As perhaps Goffman (1974, 1967, 1959) and Collins (1986a, 1985, 1975) were first to recognize and appreciate fully, Durkheim's statements on ritual have broader theoretical implications, as Durkheim himself consistently hinted (see especially Durkheim [1912] 1965, 194–298). This theory, abstracted from the context of religion, is modeled in the bottom portions of figure 7.1.

Copresence among individuals, especially when involving movement to a high-density situation (note causal arrows to the copresence variable), causes people to interact; and the higher their rates of interaction, the greater the emotional arousal (Durkheim [1912] 1965, 240–43). Conversely, as the causal arrows in figure 7.1 underscore, emotional arousal will increase rates of interaction. Copresence will also produce, directly by itself, the emission of rituals, especially under conditions in which actors move into situations of high density. However, Durkheim ([1912] 1965, 240) also indicated that rituals are more likely when rates of interaction are high and emotions are aroused—a situation he typified as "effervescence." High levels of ritual performance, Durkheim argued, feed back to increase the desire for increased interaction and to raise the level of emotion. Such a closed interpersonal system obviously cannot cycle in this mutually reinforcing manner forever—if only because of physical exhaustion—but Durkheim clearly understood the critical interpersonal dynamics producing "solidarity." Structurally, such solidarity is manifested in two ways: (1) increasing the "volume," "intensity," and "determinateness" of the "collective conscience" for the individuals involved and (2) increasing the level of subgroup formation, or density and intensity of ties among individuals. Reciprocally, high levels of "volume," "intensity," and "determinateness," coupled with high "intensity" and "density" of ties in subgroups, will feed back (in a reverse causal chain) and increase the level of ritual performance—at least to the point where the social relations are deified as an external and sacred "force." More typically, however, everyday rituals produce common sentiments and group ties that fall far short of such connotations of "sacredness."

Durkheim recognized these processes by the time he wrote the preface to the second edition of *The Division of Labor*, since here he advocates subgroup formation, coupled with an attendant subcollective conscience, as the basis for integration in differentiated societies (Durkheim [1902] 1933). At this time, however, he had not articulated the interpersonal processes—copresence, ritual, interaction, and emotional arousal—or their causal connections that produce and reproduce such subgroups.

As the model indicates, these group formation processes will produce "egoism" if copresence and interaction do not increase ritual performance, as well as "anomie" if normative agreements do not emerge as part of the "determinateness" of the collective conscience. Such conditions will escalate selection pressures for group formation through increased ritual performance, as indicated by the feedback loops on the bottom right of figure 7.1 (again, these are more like reverse causal chains than true feedback loops in a purposive, self-regulating cybernetic system).

Durkheim's Micro-Macro Linkage

While the model in figure 7.1 is too complex, and perhaps reveals too many vague causal connections, it provides a sense of how Durkheim visualized social reality at the macro- and microlevels. Moreover, the model gives us a set of proposals, if only implicitly, for how the two levels can be reconciled. Macrostructures are produced and reproduced through copresence, interaction, emotional arousal, and ritual as these create subgroups, norms, and other collective symbols. Conversely, macrostructures set the parameters for these microlevel processes by generating selective pressures associated with differentiation, patterns of structural coordination, and symbolic generalization. These pressures for activization of microlevel processes can be escalated where levels of structural disjuncture, anomie, and egoism are high.

There is, of course, only so far that we can go with this kind of micro-macro analysis, since Durkheim's conceptualization of two critical processes—power and inequality—is so weak. Because of this, I have not included power and inequality in the model, but obviously power/control and inequality/stratification need to be inserted into the model to complete the micro-macro linkages suggested in figure 7.1. This task is far beyond my intent here. Instead, my goal is to represent the strong points of Durkheim's theory, first as a causal model and next as a series of abstract laws in which causality is a secondary consideration.

DURKHEIM'S LAWS OF SOCIAL ORGANIZATION

My vision of scientific theory is deductive in this sense: (1) Formulate abstract laws that pertain to generic social processes; (2) later, derive corollaries that pertain to basic classes of empirical events; and (3) finally, develop specific hypotheses to test the plausibility of the abstract laws and corollaries. Without elaborating my views and the controversy that this positivistic argument now generates (see Turner 1987, 1985a), let me assert that such a view of theory is far closer to Durkheim's than many commentators on Durkheim appear willing to acknowledge or admit (e.g., Alexander 1982). Thus, the conversion of Durkheim's ideas into formal laws is in keeping with his underlying position on scientific sociology (Turner 1981).

The Law of Structural Differentiation

This principle was borrowed by Durkheim from Spencer, with relatively few alterations (see Turner 1985b, 1981, for more detailed commentary on this point). Let me state the law, and then offer a few discursive comments.

I. The degree of differentiation among a population of actors is a gradual *s*-function of the level of competition among these actors, with the latter variable being an additive function of
   A. the size of this population of actors,
   B. the rate of growth in this population,
   C. the extent of ecological concentration of this population, and
   D. the rate of mobility of actors in this population.

This proposition summarizes Durkheim's ([1893] 1933, 256–82) basic line of argument on the "causes" of the division of labor. I have simply raised the level of abstraction somewhat and stated the form of the relationship, without delving into causality. Is this law plausible? I think so, although it has been assessed empirically only in two literatures, organizational theory (e.g., Hannan and Freeman 1977; Meyer 1972; James and Finner 1975; Hendershot and James 1972; Blau 1970; Childers, Mayhew, and Gray 1971) and social ecology theory (e.g., Hawley 1986, 1950; Nolan 1979).

The Law of Cultural Differentiation

II. The degree of consensus over, and intensity of, cognitive orientations and regulative cultural codes among the members of a population is an inverse function of the degree of structural differentiation among actors in this population and a positive, multiplicative function of their (a) rate of interpersonal interaction, (b) level of emotional arousal, and (c) rate of ritual performance.

In this law, Durkheim's argument sees the operation of potentially contradictory, or at least intersecting, forces. Social differentiation reduces not only the degree to which actors share the same cognitive orientations (beliefs, interpretative schemes, stocks of knowledge, etc.) and regulative codes (specific normative understandings of rights and duties) but also the intensity of these orientations and codes (that is, their power to circumscribe thought and action). Structural differentiation can also decrease rates of interaction and ritual performance by partitioning actors; and thus, Durkheim implicitly (although in a rather groping way) specified the mechanism by which differentiation produces these "weakening" effects on the collective conscience—i.e., reduction in rates of interaction and solidarity-producing rituals. Yet, if rates of interaction and ritual can remain high under conditions of differentiation (as Blau [1977] suggests in his theory of "intersecting parameters"), then the culturally disintegrative effects of structural differentiation are muted. Hence, Durkheim's theory proposes two contradictory forces whose respective values determine the level of pressure for sociocultural disintegration.

The Law of Sociocultural Disintegration

III. The level of disintegrative pressures among a population of actors is a positive function of the degree of structural differentiation among members of this population and an inverse multiplicative function of their (a) rate of interaction, (b) level of emotional arousal, and (c) rate of ritual performance.

As indicated above, this law qualifies proposition 2, above, in this sense: If differentiated actors can sustain high rates of interaction and ritual, these interpersonal activities encourage common cognitive orientations and regulative codes, even among actors situated in very different locations in the structural morphology of a population. Hence, for Durkheim, there is a fundamental relationship in the social universe among differentiation, rates of interaction, and levels of ritual performance, on the one hand, and disintegrative forces, on the other. Some of the corollaries to this law suggested by figure 7.1 can, I think, incorporate Durkheim's analysis of "pathological forms" and, at the same time, specify some of those conditions influencing the values for the variables in law III.

A. The greater the level of structural differentiation among members of a population, the more likely are they to develop abstract and generalized cognitive orientations and regulative codes to bridge their differences in structural location; and the more this process occurs without a corresponding increase in normative specificity and subgroup formation sustained by ritual performances, the greater the level of anomie, and hence, the greater the level of disintegrative pressure among members of this population.

B. The greater the level of structural differentiation among members of a population and the more members of this population fail to develop specific regulative codes as reinforced by ritual performances, the greater the level of structural disjuncture in their social interdependencies, and hence, the greater the level of disintegrative pressure among members of this population.

C. The greater the level of structural differentiation among members of a population and the more members of this population fail to develop normatively regulated subgroupings that increase their rates of intragroup interaction and ritual performance, the greater the rate of egoism, and hence, the greater the level of disintegrative pressure among members of this population.

Although Durkheim viewed selection pressures for avoiding subpropositions IIIA, IIIB, and IIIC above as inevitable, I would argue that the values of the variables in these propositions are an empirical question—that is, particular historical systems have, for a wide variety of situationally specific reasons, been able to increase or decrease the values of these variables. Such historical processes are not, of course, the subject of theory, but the data to assess the plausibility of Durkheim's argument.

The Law of Sociocultural Integration

The converse of propositions III, IIIA, IIIB, and IIIC can be formulated as a law of integration—that is, of those forces increasing coordinated interrelations and group attachments among the members of a population.

IV. The degree of sociocultural integration among the members of a population is an inverse function of the degree of structural differentiation and a positive, multiplicative function of
A. the degree of consensus over cognitive orientations and regulative codes among members of this population,
B. the rate of interaction among members of this population,
C. the rate of ritual performance among members of this population,
D. the level of interdependence among members of this population, and
E. the density of group/subgroup relations or networks among members of this population.

At first blush, this law appears to be an obvious tautology, since (A), (B), (C), (D), and (E) above are the defining characteristics of sociocultural integration. The power of law IV, however, rests on the effects of structural differentiation per se and on the independent as well as multiplicative relations among the variables denoted in (A), (B), (C), (D), and (E). The values for these variables will differ depending upon empirical conditions, but at the same time, they can exponentially increase each other's effects, or perhaps cancel each other out. For example, different patterns of sociocultural integration will ensue when we compare a population revealing high structural differentiation, high levels of consensus but low intensity (i.e., generalized symbolic codes with low regulative power), low levels of interdependence among others, moderate rates of interaction and ritual, and low network densities in subgroup formation with another population that evidences high values for all the variables. Durkheim's famous distinction between mechanical and organic solidarity represented an attempt to illustrate how varying profiles of integration ensued depending upon the values of the variables in proposition IV; but my representation of Durkheim's ideas allows for more diverse patterns and profiles of integration than a dichotomous distinction between mechanical and organic solidarity. Structural differentiation, as its effects are compounded by the variety of possible interactive combinations of (A), (B), (C), (D), and (E) in proposition IV, can create a wide variety of sociocultural profiles, or patterns of integration. In this way, I think, the seeming tautology in law IV is obviated, in much the same way as plugging empirical values obviates other tautologies, or statements of equivalence in science, such as $F = ma$ (since "force" is defined in terms of mass × acceleration)—that is, as one makes deductions to specific classes of empirical cases, the tautological character of the variables is obviated.

CONCLUSION

Laws I, II, III, and IV capture, I think, the essence of Durkheim's theory. One can make many deductions from, as well as elaborations of, these four laws. For example, Merton's (1957) famous theory of deviance is, in essence, an effort to explain rates of deviance in a population by creating a corollary to laws II, IIIA, IIIC, and IV—that is, Merton conceptualizes the rate and type of deviance as a function of the particular profile among differentiation, consensus, and intensity of cognitive orientations/regulative codes and subgroup formation. To illustrate further, much of the human ecology school (Hawley 1986, 1950) in American sociology represents an implicit deduction from law I. Further, Collins's (1975) theory of interaction ritual chains is an effort to make deductions from law III. And Durkheim's ([1897] 1951) own analysis of suicide is a similar deductive explanation in terms of the values for the variables in principles IIIA (anomic suicide) and IIIC (altruistic and egoistic suicide). But my purpose here is not to make such systematic deductions for these and many other "Durkheimian" theories, especially since I am not sure that I have stated the laws in their most exhaustive and, at the same time, most succinct form. My purpose in this chapter is to suggest how we should be treating the genius of Émile Durkheim, at least from the point of view of science.

NOTES

1. Durkheim does, of course, address the dynamics of power and inequality. In his analysis of the "forced division of labor" (Durkheim ([1893] 1933, 374–88), the relationship between social class and anomie (Durkheim ([1897] 1951, 248–54), and the portrayal of political processes revolving around occupational groups ([1902] 1933) all contain interesting insights. Indeed, the discussion of class and anomie in *Suicide* clearly becomes the basis for one of the most interesting theoretical ideas of this century: Merton's (1957) famous social structure and anomie theory of deviance. Yet, despite hints and glimpses, each of which provides theoretical leads, Durkheim does not develop or extend his analysis of inequality and power in the same ways as the ideas to be analyzed in this chapter.

2. It could be argued that *the cause* of Newton's formulas is "gravitational attraction," but this is simply a gloss, because the mechanisms or processes by which gravitational attraction operates are unclear. What is it that makes gravitational attraction a force in the universe? Or, why are bodies "attracted" to each other?

3. For example, Wallace (1987) recently addressed the issue in sociology, and in the same issue of *Sociological Theory*, a "symposium" on "cause, law, and probability" was organized. Hence, the issue of cause is still very much with us and as controversial as ever.

4. For example, if a biologist or physiologist asserts that the function of the heart is to circulate blood and air to the cells, there is an implied selection argument: Among larger warm-blooded animals, those that could develop a more efficient pumping and circulatory system were more likely to survive and reproduce.

5. This statement is an illegitimate teleology, because there is no purpose/goal built into the system, such that decision makers establish "system integration" as their goal and then engage in a series of initiatives to bring about a division of labor to realize this goal.

6. Durkheim's ([1904] 1933) vision of a political system involved representative democracy in which the centralized power of the state was checked and balanced by the power of "occupational groups." Such groups would elect representatives who would assure that the power of the state would not become too great and would remain responsive to the

needs of members in diverse occupational groups, whose varied interests would serve as another check and balance on concentrated power.

7. Perhaps Durkheim had in mind a regulatory political system that was indeed self-regulating. But his statements are so moralistic that it is hard to sort out just what he meant. Hence, it is better to conceptualize these processes as reverse causal chains, although further development of Durkheim's theory would, no doubt, involve a conceptualization of power/decision-making and true feedback processes of self-regulation.

8 Marx and Simmel: Reassessing the Foundations of Conflict Theory

Over the last thirty years a number of impressive attempts have been made to uncover the "laws" of social conflict.[1] This steady progress in developing theoretical formulations has been possible because of the formative efforts of previous generations of scholars. Indeed, the sociology of conflict has stood, to borrow Newton's famous acknowledgment, "on the shoulders of giants," including Hobbes, Bodin, Smith, Mill, Malthus, Marx, and Simmel. It would be foolish, of course, to ignore the significance of an expanding research literature on the development of these theories, but perhaps more than any other theoretical tradition, conflict sociology is indebted to the conceptual accomplishments of its forefathers.

Nowhere is this historical debt more evident than in the contemporary conflict theories of Ralf Dahrendorf (1957, 1958) and Lewis Coser (1956, 1967), who have sought to expand, respectively, the Marxian and Simmelian traditions into a more viable theoretical perspective (Turner 1975). In a less explicit but no less profound way, Marx and Simmel have exerted an enormous influence on other conflict theorists. And it is probably not too much of an exaggeration to maintain that Marx's and Simmel's works have been the most influential on the sociology of conflict. In light of this influence on contemporary theory, therefore, I think it appropriate to reexamine Marx's and Simmel's contributions to the theory of

This chapter was originally titled "Marx and Simmel Revisited: Reassessing the Foundations of Conflict Theory" and appeared in *Social Forces* 53 (1975): 617–27.

conflict processes and to determine whether their schemes can still inform a sociology of conflict.

CONTRASTING ORIENTATIONS TO THE STUDY OF CONFLICT

While there are points of convergence in Marx's and Simmel's theoretical approaches to the study of conflict, the sharp differences are much more prominent. Part of the explanation for these differences resides in the diverging purposes of Marx's and Simmel's analyses. Simmel ([1908], 1956) was concerned primarily with abstracting the "forms" of social reality from ongoing social processes, whereas Marx (1848, 1867) was committed to changing social structures by altering the course of social processes. Thus, Simmel's analytical scheme was the product of a more passive and less passionate assessment of conflict, while Marx's scheme reflected political commitment to activating conflicts that would change the structure of society.

These differences in the purposes of Marx's and Simmel's studies of conflict are reflected in the assumptions they held about the nature of the social world. Marx's commitment to social change led him to visualize social systems as rife with change-producing conflict. While most of his theoretical work was concerned with revolutionary class conflicts in industrial societies, the basic assumptions underlying this substantive concern can be stated more abstractly (Turner 1973, 1974): (1) Although social systems reveal interdependence of units, these interrelations always reveal conflicts of interest. (2) These conflicts of interest are the result of the unequal distribution in all social systems of scarce resources, particularly power. (3) Latent conflicts of interest will eventually lead to overt and violent conflict among social groupings in a system. (4) Such conflicts will tend to become bipolar, since the unmasking of true interests reveals that a small minority holds power and exploits the large majority. (5) The eruption of conflict leads to social reorganization of power relations within a system. (6) This reorganization once again creates conditions of conflicting interests that set into motion inevitable processes of bipolar conflict and system reorganization.

In contrast to these assumptions are those developed by Georg Simmel, who, like Marx, saw conflict as a pervasive feature of social systems but also, unlike Marx, failed to perceive systems as typified solely by conflicting interests inherent in relations of domination and subjugation. Rather, social systems were viewed as a mingling of "associative and dissociative processes" that were more easily separated in abstract analysis than in empirical fact (Simmel [1908] 1956, 23–24). For whatever the actual nature of social structure, "we put it together, *post factum*, out of two tendencies, one monistic, the other antagonistic." Furthermore, Simmel viewed conflict as a reflection of more than just conflicts of interests, but also, of conflicts arising from hostile "instincts." Such instincts can be exacerbated by conflicts of interest, or mitigated by harmonious social relations and counteracting instincts for love, but in the end, Simmel still saw one of the ultimate sources of conflict as residing in the biological makeup of humans. For Simmel, then, systems without conflict would show "no life process," and the central theoretical task lay in discovering the basic forms of this "life process."

A concern with the forms of social interaction led Simmel to study the interplay between "associative" and "dissociative" processes and how they operate to create and maintain social patterns. Such a concern with assessing the consequences of conflict on social forms apparently dictated a set of assumptions about the nature of conflict that differed markedly from Marx's: (1) The systemic features of social systems can be typified as an organic intermingling of associative and dissociative processes. (2) These processes arise out of the instinctual impulses of actors and the imperatives dictated by various types of social relationships. (3) Conflict is thus a reflection of instinctual impulses and conflicts of interest as these are mitigated by other associative instincts and social processes. (4) While conflict is inevitable and pervasive, it does not always lead to change in social forms. (5) In fact, conflict can operate to maintain the basis of integration in social systems.

The contrasting purposes of their analyses, coupled with their diverging assumptions about the nature of conflict, are reflected in Marx's and Simmel's diverging conceptualizations of the variables in their respective theoretical schemes. Since Simmel was looking for the basic *forms* of interaction, it is more likely that he would perceive the variable properties of conflict. On the other hand, by virtue of his political commitment to rapid social change, Marx would be more likely to focus on violent conflict processes that could initiate desired social changes. Thus, while Simmel did not follow his carefully drawn analytical distinctions, he formally conceptualized the variable properties of conflict phenomena in terms of (a) the degree of regulation, (b) the degree of direct confrontation, and (c) the degree of violence between conflict parties. The end states of the ensuing variable continuum were "competition" and the "fight," with competition involving the more regulated strivings of parties toward a mutually exclusive end and with the fight denoting the more violent and unregulated combative activities of parties directly toward one another (Simmel 1908, 58). On the other hand, Marx paid little analytical attention to the variable properties of conflict processes and focused primarily upon their violent manifestations as social classes directly confront one another.

Perhaps the most significant contrast between Marx's and Simmel's orientations as they bear on the development of conflict theory is the position of conflict variables in their respective causal schemes. For Simmel, conflict was considered to "cause" various outcomes for both the social whole and its subparts. The kind of outcome or function of conflict for the systemic whole or its parts was seen by Simmel to vary with the degree of violence and the nature of the social context. Marx was also concerned with how conflict causes certain outcomes for social wholes, but unlike Simmel, he fixed attention largely on the causes of the conflict itself. Thus, for Simmel, the sources of conflict remain unanalyzed with emphasis on conflict intensity and its outcomes for different social referents, while for Marx, the variables involved in the emergence of conflict groups are given considerably more analytical attention than the variables affecting its outcomes. These differences in the position of the variables in their schemes reflect Marx's and Simmel's contrasting assumptions and purposes. Simmel perceived the source of conflict to be buried in a constellation of "associative and dissociative" processes, as well as in human "instincts of hate." Apparently, the variability and complexity of these sources of conflict made analysis too formidable. Hence, Simmel considered it more prudent to focus on the consequences of conflict once it was initiated. For in the end, what was most important for Simmel was discovering the consequences of variations on the basic forms of social interaction—a task that he apparently felt could be accomplished without delving far into human instincts. On the other hand, Marx's commitment to dialectical assumptions about conflict and change made questions of the consequences of conflict on social forms easy to answer: radical alteration of the social order. The more important issue for Marx was documenting how such change-producing conflict could emerge in the first place, with the result that Marx's conceptualization of variables focused almost exclusively on the causes of violent conflict.

These contrasts in Marx's and Simmel's purposes of analysis, assumptions about the nature of conflict, and conceptualizations of conflict variables are but necessary groundwork for what is theoretically most important: comparing their propositions on conflict processes. For in the end, a theory is only as good as the testable propositions it can generate. Analysis of theoretical orientations of scholars such as Marx and Simmel is thus useful only in that it provides information on *why* certain propositions are developed and *why* others are given little attention.

CONTRASTING THEORETICAL PROPOSITIONS

The extent to which the propositions of both Marx and Simmel underlie many current attempts at building conflict theory becomes most evident when they are stated abstractly and thereby divorced from either their polemic or their discursive context. While much of the substantive flavor of each author's discussion is lost in such an exercise, the theoretical significance of their more abstract ideas for a sociology of conflict can be made more explicit.

For the Marxian scheme in particular, I think it advisable to abstract above Marx's polemics and pull out only the most basic propositions. While this approach may offend Marxian scholars,[2] it is necessary to supplement their exhaustive and fascinating scholarship with a more succinct summary of Marx's contribution to the *theory*—as opposed to philosophy and polemics—of conflict. Thus, the first basic proposition in his scheme can be briefly stated as follows:

I. The more unequal the distribution of scarce resources in a system, the greater will be the conflict of interest between its dominant and subordinate segments.

In this proposition, Marx viewed the degree of inequality in the distribution of scarce resources, most notably power, as determining the objective conflict of interests between those with and those without power. This proposition follows directly from Marx's assumption that in all social structures, the unequal distribution of power inevitably creates a conflict of interests between superordinates holding power and subordinates lacking power. Marx's next theoretical task then involved a documentation of conditions under which a sufficiently high level of awareness of this inherent conflict of interests can cause subordinates to begin questioning the legitimacy of current patterns of resource distribution. The conditions translating awareness into a questioning of legitimacy are summarized in propositions II, IIA, IIB, IIC, and IID below:

II. The more subordinate segments become aware of their true collective interests, the more likely they are to question the legitimacy of the unequal distribution of scarce resources.
   A. The more social changes wrought by dominant segments disrupt existing relations among subordinates, the more likely are the latter to become aware of their true collective interests.
   B. The more practices of dominant segments create alienative dispositions among subordinates, the more likely are the latter to become aware of their true collective interests.
   C. The more members of subordinate segments can communicate their grievances to each other, the more likely they are to become aware of their true collective interests.
      1. The more spatial concentration of members of subordinate groups, the more likely are they to communicate their grievances.
      2. The more subordinates have access to educational media, the more diverse the means of their communication, and the more likely are they to communicate their grievances.

D. The more subordinate segments can develop unifying systems of beliefs, the more likely they are to become aware of their true collective interests.
  1. The greater the capacity to recruit or generate ideological spokespersons, the more likely ideological unification.
  2. The less the ability of dominant groups to regulate the socialization processes and communication networks in a system, the more likely ideological unification.

In these basic propositions, Marx indicated that the more dominant groups disrupt the existing relations of subordinates, thereby breaking down the very patterns of social organization that have limited the vision of subordinates, the more likely are subordinates to objectively perceive their actual situation and alternatives to their continued subordination. For as long as social relations remain stable, it is difficult for subordinates to see beyond the immediate exigencies of their existence. Disruption of life situations is likely to lead to increased awareness, especially when the activities of subordinates are highly alienating, allowing little emotional involvement and satisfaction. However, disruptive change in, and alienation from, current social relations are insufficient to cause widespread awareness of true interests: It is also necessary for subordinates to communicate, and mutually reinforce, their grievances. Such communication is more likely to occur when subordinates are in close proximity to one another and when they can become exposed to educational media, thereby liberating them from traditional means of socialization and communication. But in Marx's theory, mere communication of grievances is insufficient to cause intense questioning of legitimacy. It is also necessary for these grievances to become codified in a unifying belief system that can emphasize the *common* plight and interests of subordinates. The codification of such a belief system is most likely when ideological spokespersons, who can present a consistent viewpoint in an appealing manner, can be recruited. These spokespersons, and the emerging belief system, can be most effective when dominant groups are unable to regulate and completely control socialization processes and communication networks.

Once subordinates become aware of their common interests, the next stage in the conflict process, as Marx saw it, involves political organization to pursue conflict. These organizational processes are summarized in propositions III, IIIA, IIIB, IIIC below:

III. The more subordinate segments of a system are aware of their collective interests, the greater their questioning of the legitimacy of the distribution of scarce resources, and the more likely they are to organize and initiate overt conflict against dominant segments of a system.

A. The more the deprivations of subordinates move from an absolute to a relative basis, the more likely subordinates are to organize and initiate conflict.
B. The less the ability of dominant groups to make manifest the collective interests, the more likely are subordinate groups to organize and initiate conflict.
C. The greater the ability of subordinate groups to develop a leadership structure, the more likely they are to organize and initiate conflict.

In these propositions, Marx summarized some of the conditions leading to those forms of political organization that, in turn, will result in overt conflict. The first key question in addressing this issue is *why* an awareness of conflicting interests and a questioning of legitimacy of the system would lead to organization and the initiation of conflict. Seemingly, awareness would have to be accompanied by intense emotions if people are to run the risks of opposing those holding power. Presumably Marx's proposition on alienation would indicate one source of emotional arousal, since for Marx, alienation goes against people's basic needs. Further, ideological spokespersons would, as Marx's own career and works testify, arouse emotions through their prose and polemics. But the key variable in the Marxian scheme is "relative deprivation." The emotions aroused by alienation and ideological spokespersons are necessary but insufficient conditions for taking the risks of organizing and initiating conflict against those with power. Only when these conditions are accompanied by rapidly escalating perceptions of deprivations by subordinates is the level of emotional arousal sufficient to prompt political organization and open conflict with superordinates. Such organization, however, is not likely to be successful unless dominant groups fail to organize around their interests and unless political leaders among the subordinates can emerge to mobilize and channel aroused emotional energies.

Thus, while Marx assumed that conflict is inevitable, his theory of its causes was elaborate and set down a series of necessary and sufficient conditions for its occurrence. It is in these propositions that Marx's great contribution to a theory of conflict lies, for his subsequent propositions appear to be simple translations of his dialectical assumptions into statements of covariance without the careful documentation of the necessary and sufficient conditions that would cause these conflict processes to occur.

In his next propositions, Marx attempted to account for the degree of violence in the conflict between politically organized subordinates. The key variable was polarization, a somewhat vague concept denoting the increasing partitioning of a system into two conflict organizations:

IV. The more subordinate segments are unified by a common belief and the more developed their political leadership structure, the more dominant and subjugated segments of a system will become polarized.
 V. The more polarized the dominant and subjugated, the more violent the ensuing conflict will be.

In contrast to his previous propositions, propositions IV and V do not specify any conditions under which polarization will occur, nor do they indicate when polarized groups will engage in violent conflict. Marx just assumed that such would be the case as the dialectic mechanically unfolded. Presumably, highly organized subordinates in a state of emotional arousal will engage in violent conflict, but as a cursory review of actual events underscores, such a state often has the opposite result: less violent conflicts with a considerable degree of negotiation and compromise. This fact points to the Marxian scheme's failure to specify the conditions under which polarization first occurs and leads to violent conflict. It is not just coincidental that at this point in his scheme, Marx's predictions about class revolutions begin to go wrong. Thus, the Marxian legacy points rather dramatically to a needed area of theoretical and empirical research: Under what conditions is conflict likely to be violent? And more specifically, under what conditions is conflict involving *highly organized and mobilized subordinates* likely to be violent, and under what conditions are less combative forms of conflict likely to occur?

The final proposition in the Marxian inventory also appears to follow more from a philosophical commitment to the dialectic than from carefully reasoned conclusions:

VI. The more violent the conflict, the greater will be the structural change in the system and the redistribution of scarce resources.

This proposition reveals Marx's faith in the success of the revolution as well as his assertion that new sets of super-subordinate relations of power would be established by these successful revolutionaries. As such, the proposition is ideology rephrased in the language of theory, especially since no conditional statements are offered on just when violent conflict leads to change and redistribution and just when it does not. Had Marx not *assumed* conflicts to become polarized and violent, then he would have paid more attention to the *degrees* of violence and nonviolence in the conflict process, and this in turn, would have alerted him to the variable outcomes of conflict for social systems. In fact, as suggestive as Marx's propositions are, the entire scheme suffers from his failure to specify clearly the interaction of variables. For example, the scheme begs questions like these: What *kinds* or *types* of inequality create *what types* of conflict of interest? What *types* of awareness and questioning actually lead to *what degrees* of overt violence, and what *types* of ideological unification and political leadership producing what *types* of polarization leading to what *types* of violent conflict cause what *types* of structural change? It is to answering these questions in Marx's theory that contemporary social theorists must address their efforts.

In contrast to Marx's inventory of basic propositions, Simmel's offered no propositions on the ultimate causes of conflict. Rather, Simmel's propositions focus on the intensity or degree of violence or combativeness of conflict once initiated and on the consequences of conflicts for the parties to the conflict and for the systemic whole. While it is regrettable that Simmel chose not to examine the causes of conflict, his emphasis on the varying outcomes of conflict does provide some necessary corrections for Marx's scheme. For I think it is clear that Marx's inventory begins to break down at just this point in his analysis, since for Marx conflict among organized groups pursuing divergent interests must be violent and lead to dramatic social reorganization. Simmel's propositions offer some clues as to where Marx went wrong in these presumptions; and in so doing, they help recast the foundations of conflict sociology.

Simmel's propositions are not always easy to abstract from his rambling prose, especially since he tends to argue by example and analogy. As a result of this type of exposition, he constantly shifts the units and levels of analysis—from intrafamily conflict to wars between nation-states. To appreciate Simmel's significance for a theory of conflict, then, I think it best to abstract above his discursive prose and thereby present only what appear to be the most generic propositions.[3] Simmel's primary concern in analyzing the forms of dissociation in social systems was with the degree of combativeness or violence of conflict. I have summarized these basic propositions below:

I. The greater the degree of emotional involvement of parties to a conflict, the more likely is the conflict to be violent.
  A. The greater the respective solidarity among members of conflict parties, the greater is the degree of their emotional involvement.
  B. The greater the previous harmony between members of conflict parties, the greater is the degree of their emotional involvement.
II. The more conflict is perceived by members of conflict groups to transcend individual aims and interests, the more likely is the conflict to be violent.
III. The more conflict is a means to an end, the less likely is the conflict to be violent.

Propositions I, IA, and IB overlap somewhat with those developed by Marx. In a vein similar to Marx's, Simmel emphasized that violent conflict is the result of emotional arousal. Such arousal is particularly likely when conflict groups possess a great deal of internal solidarity and when these conflict groups emerge out of previously harmonious relations. Marx postulated a similar process in his contention that polarization of groups previously involved in social relations (albeit exploitive ones) leads to violent conflict. In proposition II, Simmel indicated that, coupled with emotional arousal, the extent to which members see the conflict as transcending their individual aims increases the likelihood of violent conflict. Marx analyzed, of course, in considerably more detail just how such a state of consciousness is created, since his concern with erasing false consciousness through communication and creation of a common belief system represents a more precise way of stating Simmel's proposition.

Proposition III is Simmel's most important, because it appears to contradict Marx's hypothesis that objective consciousness of interests will lead to organization for violent conflict. In this proposition, Simmel argued that the more clearly articulated their interests, the more focused are the goals of conflict groups. With clearly articulated goals, it becomes possible to view violent conflict as only one of many means for their achievement, since other, less combative conflicts, such as bargaining and compromise, can often serve to meet the now limited objectives of the group. Thus for Simmel, consciousness of common interests (Simmel's proposition II above) can, under unspecified conditions, lead to highly instrumental and nonviolent conflict. Marx's analysis precludes this possibility;[4] and while Simmel's propositions leave many questions unanswered, they do provide a corrective to the Marxian analysis: Conflict among highly organized groups of emotionally aroused actors pursuing collective goals can, under conditions that need to be specified, lead to using violent conflict as only one means to an end that, under other specific conditions, can actually lower the probability of violent conflict. In the context of labor-management relations, for example, Simmel's proposition appears to have been more accurate than Marx's, since violence has more often accompanied labor-management disputes, especially in the initial formation of unions, when interests and goals are not well articulated. As interests become clarified, violent conflict has been increasingly replaced by less violent forms of social interaction.

Thus, both Simmel and Marx provided an interesting set of propositions on how conflict groups become organized and mobilized to pursue violent conflict, but as Simmel seemed to point out, this very process of mobilization and organization can, in the end, foster less violent forms of conflict. Violence appears to be an interim result of initial organization and mobilization, but as groups become highly organized, they become more instrumental, thereby decreasing the probability of violent conflict.

The curvilinear nature of the conflict process is further clarified by Simmel's subsequent attention to the consequences or functions of conflict for the conflict parties and for the systemic whole within which the conflict occurs. For Simmel first analyzed how violent conflicts *increase* solidarity and internal organization of the conflict parties, but when he shifted to an analysis of the functions of conflict for the social whole, he drew attention primarily to the fact that conflict promotes system integration and adaptation. How can violent conflicts promoting increasing organization and solidarity of the conflict groups suddenly have these positive functions for the systemic whole in which the conflict occurs? For Marx, such a process was seen to lead to polarization of conflict groups and then to *the* violent conflicts that would radically alter the systemic whole. But for Simmel, the organization of conflict groups enables them to realize many of their goals without overt violence (but perhaps with a covert threat of violence), and such partial realization of clearly defined goals cuts down internal system tension, and hence promotes integration.

To document this argument, I have first listed below Simmel's key propositions on how violent conflicts can increase the organization of conflict parties:

I. The more violent intergroup hostilities and the more frequent conflict among groups, the less likely are group boundaries to disappear.
II. The more violent the conflict and the less integrated the group, the more likely is despotic centralization of conflict groups.
III. The more violent the conflict, the greater will be the internal solidarity of conflict groups.
    A. The more violent the conflict, and the smaller the conflict groups, the greater will be their internal solidarity.
        1. The more violent the conflict and the smaller the conflict groups, the less will be the tolerance of deviance and dissent in each group.
    B. The more violent the conflict, and the more a group represents a minority position in a system, the greater will be the internal solidarity of the group.
    C. The more violent the conflict, and the more a group is engaged purely in self-defense, the greater will be the internal solidarity.

In these propositions, violent conflict, under varying conditions, will lead to clearer definition of the boundaries of the conflict groups, centralization of the groups, and increases in internal solidarity of the groups. When viewed in the narrow context, most of these propositions overlap Marx's, but they diverge considerably when their place in Simmel's overall propositional inventory is examined. Unlike Marx's inventory, Simmel's does not assume that conflict begets increasingly violent conflicts between increasingly polarized segments in a system, which, in the end, will cause radical change in the system. This difference between Marx's and Simmel's analyses is dramatically exposed when Simmel's propositions on the consequences of conflict for the systemic whole are reviewed. The most notable feature of several key propositions listed below is that Simmel was concerned with less violent conflicts and with their integrative functions for the social whole:

I. The less violent the conflict between groups of different degrees of power in a system, the more likely is the conflict to have integrative consequences for the social whole.
    A. The less violent and more frequent the conflict, the more likely is the conflict to have integrative consequences for the social whole.
        1. The less violent and more frequent the conflict, the more members of subordinate groups can release hostilities and have a sense of control over their destiny, and thereby maintain the integration of the social whole.
        2. The less violent and more frequent the conflict, the more likely are norms regularizing the conflict to be created by the conflict parties.
    B. The less violent the conflict and the more the social whole is based on functional interdependence, the more likely is the conflict to have integrative consequences for the social whole.
        1. The less violent the conflict in systems with high degrees of functional interdependence, the more likely it is to encourage the creation of norms regularizing the conflict.

These propositions provide an important qualification of Marx's analysis, since Marx visualized mild conflicts between super- and subordinate as intensifying as the conflict groups become increasingly polarized; and in the end, the resulting violent conflict would lead to radical social change in the system. In contrast, Simmel argued that conflicts of low intensity and high frequency in systems of high degrees of interdependence do not necessarily intensify or lead to radical social change. On the contrary, they release tensions and become normatively regulated, thereby promoting stability in social systems. Further, Simmel's previous propositions on violent conflicts suggest the possibility that with the increasing organization of the conflicting groups, the degree of violence of their conflict will decrease as their goals become better articulated. The end result of such organization and articulation of interests will be a greater disposition to initiate milder forms of conflict, involving competition, bargaining, and compromise. What is critical for developing a sociology of conflict is that Simmel's analysis provides more options than do Marx's propositions on conflict outcomes. First, conflicts do not necessarily intensify to the point of violence, and when they do not, they can have, under conditions that need to be further explored, integrative outcomes for the social whole. Marx's analysis precludes exploration of these processes. Second, Simmel's propositions allow for inquiry into the conditions under which initially violent conflicts can become less intense and thereby have integrative consequences for the social whole. This insight dictates a search for the conditions under which the level of conflict violence and its consequences for system parts and the social whole can shift and change over the course of the conflict process. This expansion of options represents a much broader, and I suspect firmer, foundation for building a theory of conflict.

Finally, Simmel presents two basic propositions on the positive functions of violent conflict for expanding the basis of integration of systemic wholes:

III. The more violent and prolonged the conflict relations between groups, the more likely is the formation of coalitions among previously unrelated groups in a system.

IV. The more prolonged the threat of violent conflict between groups, the more enduring are the coalitions of each of the conflict parties.

These propositions could represent a somewhat different statement of Marx's polarization hypothesis, since conflict was seen by Simmel as drawing together diverse elements in a system as their respective interests become more clearly recognized. But Simmel was not committed to dialectical assumptions, and thus, he appeared to be arguing only that violent conflicts pose threats to many system units, which, depending upon calculations of their diverse interests, will unite to form larger social wholes. Such unification will persist as long as the threat of violent conflict remains. Should violent conflict no longer be seen as necessary, with increasing articulation of interests and the initiation of bargaining relations, then Simmel's propositions IA and IB on the consequences of conflict for the social whole would become operative. Thus once again, Simmel's analysis on conflict offers more options in developing a theory of how varying types of conflict can have diverse outcomes for different system referents at different points in the conflict process.

CONCLUDING COMMENT

In this chapter, I have tried to present the key elements in Marx's and Simmel's theories of conflict. These theories are most useful when stated at their most abstract levels, for it is in this form that the debt of contemporary theorizing to these two German scholars becomes most evident. For those who would resent such a brief and concise overview, I can only offer the belief that it is time to pull from these and other scholars what is theoretically most useful and move on with the job of theory building.

Such an exercise as this becomes possible only with the intellectual perspective provided by the previous scholarship of others. Yet, theory cannot be the history of ideas, nor should it involve only debates as to which scholar said what. In the end, theory must be bodies of interrelated and abstract statements relating clearly defined variables. Thus, while this paper lacks the typical scholarship of commentaries on these thinkers, it does take Marx and Simmel seriously as social theorists and therefore asks: What are the interrelated and abstract propositions of their theories? I have sought to provide my tentative answer to this most important question and would encourage corrections to my interpretation of Marx's and Simmel's concepts and propositions. In this way, we can begin to appreciate the significance of these intellectual giants as true theorists in the science of sociology.

NOTES

1. The list of theoretical efforts is long, but representative examples would include the works of Williams (1947, 1970), Mack and Snyder (1957), Schelling (1960), Boulding (1962), Blalock 1967, and Collins (1975).

2. Probably more offensive to Marxian scholars is my firm conviction, after my own intense reading of Marx over a number of years, that Marx's theory is most explicit in his polemical essays. And further, *the best* statement of his theory came early in *The Communist Manifesto* (1848). Although Marx was to have second thoughts near the end of his career on some of the hypotheses contained in this work, these thoughts were never formulated into clear theoretical statements. For those who would be shocked at boiling down Marx's thought to a few propositions, I offer solace in the fact that these few propositions have been sufficiently profound to shape the course of conflict theory in contemporary sociology.

3. These propositions are abstracted from Simmel's essay on conflict in Kurt H. Wolff's translation of *Conflict and the Web of Group Affiliations* (1956). The propositional inventory presented here and elsewhere differs considerably from Coser's, primarily because many of what Coser (1956) chose to call "propositions" are, in my view, definitions or assumptions. I have also taken more liberty than Coser to rephrase and state more generically Simmel's propositions. Further, in some instances I have omitted propositions that are not critical to the basic argument. For a complete listing of Simmel's propositions, see Coser (1956) and Turner (1974).

4. Admittedly, Marx's late awareness of the union movement in the United States forced him to begin pondering this possibility, but he did not incorporate this insight into his theoretical scheme.

9 Where Marx Went Wrong

At a time when "Marxism" as a sociopolitical movement is receding and its archenemy, capitalism, is ascendant, it would be easy to call into question the whole Marxian project. But if our goal is science, as opposed to political advocacy, it is not relevant that a political doctrine and set of programs drawn rather loosely from this doctrine have gone awry. Rather than address the question of why Marxism went wrong, my concern is why the underlying theory developed by Marx did not accurately predict the flow of empirical events, especially at the very beginning of the revolutionary movements that ushered in communism. In particular, I will focus on Marx's theory of conflict in systems of inequality, for when all the political dust has settled, it is this theory that will mark Marx's enduring contribution to scientific sociology.

While the analysis of capitalism as a social form in *Das Capital* (Marx [1867] 1967) is truly profound, it is not a theory, but a description of how capitalism emerged and operates. It has theoretical content, but it would be difficult to generalize beyond the specific historical epoch examined by Marx. Of course, Marx never believed that general and universal theory with timeless laws is possible, so my comments here are hardly a devastating criticism. But the fact that "the revolution" did not occur as predicted and with the selected protagonist is a relevant comment and criticism. The reason for Marx's failings is that his theory of conflict was stated in one of his most polemical works, *The Communist Manifesto* (Marx and Engels [1848] 1932); and these polemics, and the wishful thinking in them, led him to make certain key theoretical mistakes. But much of the theory, as an abstract statement of the conditions producing conflict in systems of inequality, is still useful, once the mistakes are highlighted and corrected.

As I explored from a somewhat different angle in the last chapter, when Karl Marx's and Georg Simmel's theories of conflict are compared it becomes evident that the basic problem in Marx's theory is this: He assumed that high degrees of political and ideological mobilization of actors into conflict groups would lead to a violent and change-producing revolution. Such is infrequently the case, for as parties to a conflict become well organized and unified, they also clarify their goals and the costs involved in realizing them, with the result that they are more likely to bargain and compromise. This is my argument, which mirrors Simmel's in its broad contours; now, let me fill in the details with a more careful analysis of Marx's theory of conflict.

THE THEORY OF CONFLICT

In table 9.1, I have again listed Marx's propositions for easy reference. The basic argument is as follows: Inequalities produce conflict of interests; the more subordinates in this system of inequality become aware of their interests, the more they question the legitimacy of the distribution of resources and the more they become mobilized to pursue conflict; and the more mobilized they become, the more violent the conflict, and the greater the degree of structural change that will be produced by the conflict.

   Propositions IIA–D in table 9.1 state some of the basic forces mobilizing subordinates to become aware of their interests and to question the legitimacy of the existing system of resource distribution; and coupled with the processes listed in propositions IIIA–C, additional forces push subordinates to pursue conflict with superordinates. It is in the transition from the conditions listed under proposition II to those under proposition III that Marx goes wrong. He saw—he *wanted* to see—an inevitable polarization of superordinates and subordinates who would join in conflict and transform the exploitive nature of society. But the events in propositions II and III are not cumulative and additive in quite the way visualized by Marx. Indeed, violence and structural change are much more likely to break out early in this process, usually before political leadership (IIIC) and clear ideologies (IID) have been articulated. One way to recast Marx's argument is shown in figure 9.1.

   In Marx's reasoning, every direct causal path, moving from the left to the right portions of the model, is positive and culminates in intense conflict between superordinate and subordinate resource holders (the variable on the far right of the model). But in the causal arrows leading to "intensity of conflict," the positive signs have been changed to positive curvilinear signs in three cases (+ /–, initially positive and then negative). These small changes more accurately reflect empirical reality and show where Marx went wrong. Let me discuss the model in more detail before elaborating on this conclusion.

TABLE 9.1
Marx's Basic Principles of Conflict in Systems of Inequality

I. The more unequal the distributions of scarce resources in a system, the greater will be the conflict of interests between its dominant and subordinate segments.

II. The more the subordinate segments become aware of their true collective interests, the more likely they are to question the legitimacy of the unequal distribution of scarce resources.

   A. The more the social changes wrought by dominant segments disrupt existing relations among subordinates, the more likely are the latter to become aware of their true collective interests.

   B. The more the practices of dominant segments create alienative dispositions among subordinates, the more likely are the latter to become aware of their true collective interests.

   C. The more the members of subordinate segments can communicate their grievances to each other, the more likely they are to become aware of their true collective interests.

      1. The more spatial concentration of members of subordinate groups, the more likely they are to communicate their grievances.

      2. The more the subordinates have access to educational media, the more diverse the means of their communication, and the more likely they are to communicate their grievances.

   D. The more the subordinate segments develop unifying systems of beliefs, the more likely they are to become aware of their true collective interests.

      1. The greater the capacity to recruit or generate ideological spokespersons, the more likely is an ideological unification.

      2. The less the ability of dominant groups to regulate the socialization processes and communication networks in a system, the more likely is the ideological unification of subordinates.

III. The more the subordinate segments of a system are aware of their collective interests, the greater their questioning of the legitimacy of the distribution of scarce resources, and the more likely they are to organize and initiate overt conflict against dominant segments of a system.

   A. The more the deprivations of subordinates move from an absolute to a relative basis, the more likely the subordinates are to organize and initiate conflict.

   B. The less the ability of dominant groups to manifest their collective interests, the more likely subordinate groups are to organize and initiate conflict.

   C. The greater the ability of subordinate groups to develop a leadership structure, the more likely they are to organize and initiate conflict.

IV. The more the subordinate segments are unified by a common belief and the more their political leadership structure is developed, the more the dominant and subjugated segments of the system will become polarized.

V. The more polarized the dominant and subjugated, the more violent the ensuring conflict will be.

VI. The more violent the conflict, the greater the structural change of the system and the redistribution of scarce resources.

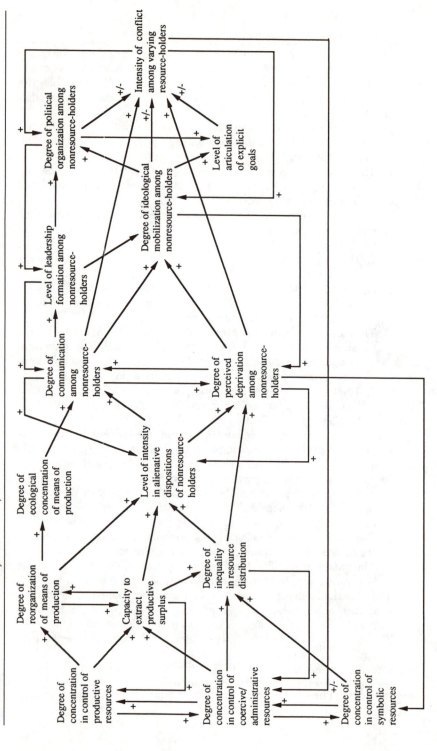

FIGURE 9.1. Marx's Theory of Conflict, Revised

Marx correctly perceived that those who control production also have disproportionate power; in turn, they use control of production and power to extract surplus and, thereby, create and sustain inequality. In addition, they use their power to control ideological resources that legitimate inequality. Marx assumed that these causal relations, as delineated at the far left of the model, were always mutually reinforcing and would set into motion a series of processes culminating in change-producing conflict. Moreover, he correctly perceived that once one of the values for these variables decreases, the positive causal connections work to transform the level of inequality. For example, a dramatic decrease in ideological control operates through the positive causal arrows to lower the values of the other variables, control of power and production, resource extraction, and inequality. This is why, presumably, he placed so much emphasis on ideological mobilization (which, as the negative reverse causal loop emphasizes, reduces the concentration of control of symbolic resources). But what Marx did not seem to appreciate is that, through the direct, indirect, and reverse causal paths portrayed in the model, different values for variables and varying configurations of reverse causal processes could interrupt and arrest the march toward intense and violent conflict. To pursue the above example, decreased concentration of ideological resources sets into motion a very complex process that can reduce the values for some variables and, hence, the potential for further ideological mobilization and conflict. Marx never really considered this possibility, at least not when speaking ideologically. But most social processes are not purely cumulative. They involve different loadings and values for the variables, and the configurations of direct, indirect, and reverse causal processes give the empirical world a more quilted and episodic character. Thus, the processes outlined in the model could work just as Marx thought, up to ideological mobilization, and then suddenly decrease as a result of the one negative reverse causal chain from ideological mobilization to concentration of symbolic resources, with the result that intense conflict never occurs.

My point here is that models like the one in figure 9.1 provide a sense of dynamics that cannot be captured in ideological traits or prose, or even in formal propositions like those listed in table 9.1. But having said this, I think that Marx was essentially correct in his analysis through most of the model: Concentrations of resources, high levels of inequality, and reorganization of production increase ecological concentration, alienation, communication, and deprivation; and these operate, via various causal paths, to increase leadership, ideological mobilization, political organization, and articulation of goals. Once these processes are set into motion, however, the reverberated effects add complications not visualized by Marx. The relationships among these latter variables and conflict do not operate as Marx theorized. The result is that all other relationships in the model are altered in ways that make Marx's predictions incorrect. Let me present this conclusion in more detail.

REASSESSING THE THEORY

The causal arrows on the right portion of the model indicate that many of the forces most directly affecting the level of intensity of conflict between subordinates and superordinates are positively curvilinear (+ /–) in that they initially work to increase conflict intensity, but as their values become high, they decrease the likelihood of violence in favor of bargaining and compromise. Moreover, as the positive feedback or reverse causal loops from conflict intensity indicate, conflict itself works via a variety of reverse causal effects to create high levels of political organization, articulation of goals, ideological mobilization, and formation of leadership that reduce conflict (by virtue of increasing the values for those variables that will, with high loadings, decrease intensity of conflict in favor of bargaining and compromise). Thus, when subordinates are ideologically mobilized, well-organized politically, and clear on their goals, violence is less likely, whereas when they are only beginning the process of ideological mobilization, political organization, and goal articulation (under the effects of the variables in the middle of the model), violence is more likely. But as violence produces more clear-cut goals and political organization, it sets into motion processes—high levels of leadership, ideological mobilization, and articulation of goals—that will generally decrease violence. Moreover, violence tends to increase the concentration of power (note the long causal loop from conflict intensity to concentrated power) and all of the forces it sets into motion—at least initially. But if conflict is of very high intensity and, I would add, of long duration, it lowers the concentration of power (hence, the + /–, or positive curvilinear relation) and initiates (through the reverberated effects along the positive causal paths) a reduction in the values of those variables leading to conflict. If these variables lower to moderate levels the variables directly connected to the intensity of conflict, violent conflict may erupt. But if the values of these variables are low, conflict may be reduced. Such are the complexities of curvilinear effects in complex, dynamic models.

This curvilinear relationship between intensity and concentrated power adds another consideration. If conflict rapidly reaches high intensity, its effects on concentrated power can also work rapidly, causing an initial mobilization of power and then a sudden collapse of coercive efforts. Whether this occurs, of course, depends on the respective resources of the antagonists and is not, as Marx hoped, inevitable. When this cycle occurs, a system can collapse politically (and economically and ideologically) before there is clear articulation of goals, consensus in ideology, and unambiguous political leadership. The result is further conflict among factions (initiating an escalating set of cycles delineated in the model) until these matters are resolved (as leaders, organization, ideology, and goals are sorted out) or imposed by a faction that can gain control of coercive capacity and, hence, of symbolic and productive resources.

CONCLUSION

The likelihood of conflict, and its intensity when it does occur, depend on a complex and highly dynamic set of processes, particularly curvilinear processes that feed back and affect the values of the variables determining the flow of the processes themselves. For Marx, who was an ideologue, this complexity would make for poor reading; instead, he had to show that a "great spectre" was haunting the capitalist world and that conflict was building to a crescendo. But the empirical world rarely accommodates these kinds of extreme arguments.

The reasons given for this conclusion are often assertions about the contingency of empirical events and the futility of believing that laws—even laws confined to an historical epoch, like capitalism—are possible. My alternative is to view processes as complex and dynamic in several senses: First, the values of variables denoting key elements of social processes vary and are often contingent in ways that cannot be predicted. Second, if the loadings of variables can differ widely in the particulars of diverse empirical circumstances, then their effects on one another—direct, indirect, and reverse—will also differ, leading to varying predictions about what is likely to occur. And third, the configurations of effects among forces, whose effects on each other are complicated by curvilinear relations, make predictions hazardous, because it is often difficult to know the reverberated consequences of complex interactions among social processes.

Such is certainly the case with conflict. But all of these complexities do not mean that positivistic theory is futile. Models like that in figure 9.1 can give us a handle on the complexity involved and allow us to delineate the way those interconnected forces operate to produce conflict. Since sociologists have not often used such models—what I termed "analytical models" in chapter 2—it is hard to know how much predictive uncertainty they can eliminate. Prediction in natural empirical systems is difficult, even when reasonably isomorphic and sophisticated models exist. Indeed, few would throw up their hands in seismic geology or meteorology because they cannot predict, respectively, an earthquake or a hurricane.

The same should be true of sociology. Our predictions will always be flawed, but the really important question is this: Does our theory, whether as a series of propositions or as a dynamic analytical model, allow us to understand *how* and *why* an event occurred, even if this understanding must be generated ex post facto? Marx cannot be faulted for flawed predictions; rather, I fault him for a theory that goes only part way in allowing us to understand how and why conflict occurs in systems of inequality.

10 Max Weber's Theory of Integration and Conflict

Because Max Weber did not advocate a scientific sociology that seeks to articulate general laws, relatively few have made the effort to extract generalizations from Weber's work. But Weber (1949) clearly had an ambiguous attitude toward science: On the one hand, he wanted sociological inquiry to be value-free and objective, whereas on the other, he remained convinced that the flow of social events is contextual, historically specific and unique, and subject to chance events. His construction of ideal types for comparing diverse empirical cases represents the outcome of this ambivalence, for an ideal type provides an objective yardstick for comparing the contextual manifestations of those social processes and forms denoted by the ideal type.

Yet, having acknowledged that Weber himself might disparage the enterprise, I believe that one of the reasons that sociologists still read and reread Weber with an almost religious fervor is because we sense that he articulated some basic laws of human organization. With some notable exceptions (e.g., Collins 1975, 1986b), few want to admit or acknowledge this possibility, but Weber's laws are nonetheless there to be observed. Indeed, Weberian laws simply pop out and hit you between the eyes, *if* you are looking for them.

To illustrate this potential in Weber's work, I will examine his theory of conflict and integration. This theory revolves around the assumption that patterns of social organization remain integrated by virtue of the level and nature of legitimacy attached to centers of political authority and that the potential for conflict and disintegration increases when legitimacy is undermined. I have, of course, already framed the issues in more positivistic language than Weber would deem appropriate, but the goal of this chapter is to illuminate those conditions articulated by Weber leading to a decline in political legitimacy and, potentially, to disintegration of social systems. In articulating these conditions, Weber also indicates how it is that social systems remain integrated. For Weber, then, conflict and integration are opposite outcomes of the same fundamental process: legitimation and, conversely, delegitimation.

THE THEORY OF ENDOGENOUS CONFLICT

It is generally agreed that Weber was engaged in a silent dialogue with Marx (Bendix 1960), but his theory of conflict and disintegration has many points in common with Marx's. In chapter 11, I will pursue this idea in more detail, but for the present, let me outline in highly abstract terms Weber's theory of integration and conflict.

For Weber, the basic organizational unit is the "legitimated order" of organized relations among variously oriented actors (in terms of Weber's ideal type on the forms of action: traditional, affectual, rational, and value-rational). My sense is that Weber typified legitimated orders with respect to the relative proportions of action orientations. A legitimated order involves a pattern of "domination"; and so, the nature of the domination determines the basis—tradition, affect, rationality—for regulating the actions of actors. Domination involves authority, or the rights of some to control and regulate the actions of others; and hence, a legitimated order involves the use of authority that is legitimated in terms of traditional, affectual, or rational criteria.

The transition from this more general discussion of social organization to Weber's famous analysis of class, status, and party is not smooth. At an organizational level, status groups and parties appear to be "communal" groupings of actors oriented in terms of some ratio of tradition, affect, and value-rationality, whereas classes are more loosely organized networks or "distributional" groupings of actors organized in terms of their rational life chances in the market. What is implied, I think, is that the system of classes, status groups, and parties constitutes a megalegitimated order for a society. This system is composed of (a) organized units, such as status groups and parties; (b) categories and loose networks of individuals marked by their chances and opportunities in the economy; and (c) rankings of (a) and (b) in a hierarchy. Each of the elements of this system—classes, status group, and parties—would appear to have an internal basis for legitimation in affect, rationality, and tradition; and moreover, each distinctive hierarchy of classes, status groups, and parties also appears to require legitimation of the domination by the higher ranks in each hierarchy in terms of some combination of affectual, rational, and traditional criteria; and finally, the overall system of hierarchies would require further legitimation of the patterns of domination by high-ranking parties, status groups, and classes.

None of these complexities are spelled out by Weber, but the general image that emerges is of a multidimensional system of rankings in terms of power, honor, and money, with each of these three hierarchies consisting of organizational units employing various criteria for action and interaction and with each dominated by particular parties, status groups, and classes. A social system remains integrated, Weber implicitly argues, when these patterns of domination within and between hierarchies of status groups, parties, and classes are considered appropriate, or legitimate, by actors. The key question thus becomes: What causes delegitimation of this system of hierarchies?

Weber's ([1922] 1968, 901–1372) answer is explicit and invokes three conditions creating the potential for delegitimation: (1) If there is a high degree of inequality and discontinuity in the rankings of status groups, parties, and classes (that is, some organizational units in the rankings have great honor, power, and material well-being, while others have very little), the potential for delegitimation increases; (2) if there is a high correlation of membership among those low and high in the status, political, and class hierarchies, the possibilities for delegitimation increase; and (3) if there are low rates of mobility among organizational units in each hierarchy and between hierarchies, the chances for delegitimation are increased even further. These ideas sound very similar to Marx's notions of immiseration and polarization, but they are presented against a more complex vision of general patterns of social organization and the system of stratification within these general patterns.

Moreover, unlike Marx, Weber did not see delegitimation as necessarily leading to conflict and revolt between those high and low in these hierarchies. The three conditions listed above simply increase the likelihood that a charismatic leader will emerge, but situationally contingent and contextual forces always operate to increase or decrease the probability of a charismatic challenger. Moreover, the ability of such a leader to be effective in articulating grievances and mobilizing resources is also contingent. Thus, "chance events" are an important part of Weber's theory of delegitimation and conflict. In figure 10.1, I have outlined the implicit Weberian model of endogenous conflict.

In the model presented in figure 10.1, legitimation tends to increase those three conditions producing delegitimation. Weber never quite *says* this, but he offers hints and suggestions that legitimation allows those who dominate to consolidate their capacity for domination; and the outcome of this process is an increase in (1) the correlation of positions in classes, status

FIGURE 10.1. Weber's Theory of Delegitimation and Conflict

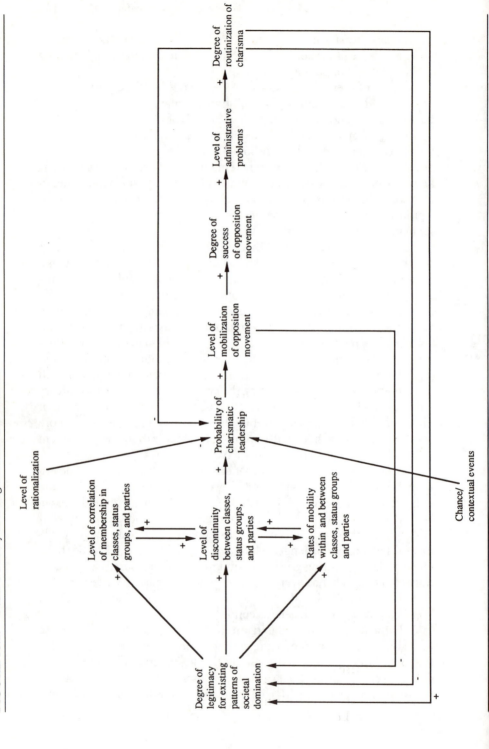

groups, and parties; (2) the level of inequality between and among the organizational units of classes, parties, and status groups; and (3) the strength of the barriers to mobility between or within classes, status groups, and parties. If situational conditions are favorable, charismatic leaders may emerge; and if contextual conditions remain favorable, a leader may prove capable of mobilizing an opposition movement; and once again, if conditions are favorable, this movement may succeed. Weber implies that this sequence of charisma, mobilization, and success is very difficult in highly rationalized systems in which domination by organizational units is legitimated in terms of rational-legal authority. Thus, charismatic leadership is most likely, he argues, when traditional authority dominates.

The feedback loops, or reverse causal chains (Stinchcombe 1968), in the model in figure 10.1 are crucial to appreciating Weber's argument. Charismatic leadership, which is based upon "personal qualities" and "affective orientations," works to delegitimate traditional authority; and as mobilization of opposition occurs under charismatic leadership, legitimacy is further eroded. This delegitimation works to arrest further consolidation of inequality, but if the movement is unsuccessful, it will reinforce the older form of legitimation and work to consolidate the stratification system around the existing pattern of rankings and resource distribution among classes, status groups, and parties. Such consolidation, however, will increase the probability that, under favorable empirical conditions in the future, yet another charismatic leader will emerge and mobilize actors in an opposition movement. Hence, for Weber legitimation sustains continuity in the organization of a population, but it also systematically generates those processes in the stratification system that can cause changes in these patterns of organization.

The "routinization of charisma" is inevitable, since the personal qualities that mobilize opposition are less useful in organizing the administration of a legitimated order. Such routinization creates a new order which, in turn, establishes a new basis of stratification revolving around hierarchies of classes, status groups, and parties. And if the reinstitution of a legitimated order is based upon tradition (and the remnants of affect), then the new stratification system can serve as the impetus behind yet another cycle of charisma, opposition, and routinization.

To highlight the positivistic strains in Weber's thinking, let me summarize his argument as a series of abstract principles. These are presented in table 10.1. When stated this way, the propositions on conflict look similar to those developed by Marx (as will be evident in chapter 11). In addition to these principles on how stratification processes work to delegitimate an order, Weber ([1922] 1968, 901–20) also developed a theory of geopolitics that provides further insight into the dynamics of integration and change. Let me now examine this theory.

TABLE 10.1
Weber's Propositions on Inequality and Conflict

   I. The greater the degree of withdrawal of legitimacy from political authority, the more likely is conflict between superordinates and subordinates.
   A. The greater the correlation of membership in class, status group, and party (or, alternatively, access to power, wealth, and prestige), the more intense the level of resentment among those denied membership (or access) and, hence, the more likely they are to withdraw legitimacy.
   B. The greater the discontinuity in social hierarchies, the more intense the level of resentment among those low in the hierarchies and, hence, the more likely they are to withdraw legitimacy.
   C. The lower the rates of mobility up social hierarchies of power, prestige, and wealth, the more intense the level of resentment among those denied opportunities and, hence, the more likely they are to withdraw legitimacy.
  II. The more charismatic leaders can emerge to mobilize resentments of subordinates in a system, the greater will be the level of conflict between superordinates and subordinates.
   A. The more conditions IA, IB, and IC are met, the more likely the emergence of charismatic leadership.
 III. The more effective charismatic leaders are in mobilizing subordinates in successful conflict, the greater the pressures to routinize authority through the creation of a system of rules and administrative authority.
  IV. The more a system of rules and administrative authority increases conditions IA, IB, and IC, the greater will be the withdrawal of legitimacy from political authority and the more likely is conflict between superordinates and subordinates.

THE THEORY OF GEOPOLITICS

In table 10.2, I have summarized Weber's principles on geopolitics and conflict in a way that connects them to those on the endogenous sources of conflict. What these propositions emphasize is that legitimation is tied to the level of prestige that a "political community" can maintain vis-à-vis other such communities. Prestige in the external geopolitical arena works to bolster legitimacy that, in turn, militates against the opposition-producing effects of stratification.

Prestige in the "world system" can be maintained in several ways: success at military conquest and economic co-optation. When economic units (e.g. chartered corporations) depend upon political authority for their viability, they encourage military conquest, whereas when economic units do not depend upon government, they usually encourage political authority to form economic alliances and co-optive arrangements. The greater the success of political authority in these activities, the greater is the level of prestige enjoyed by political authority, and hence, the greater its level of legitimation.

Legitimation is also sustained by processes revolving around external and internal threats. If external enemies exist, or can be found or manufactured, then political authority can legitimate its domination and use of power. Similarly, if internal enemies can be found, especially highly visible minority subpopulations, then the use of power to deal with this threat and with other arenas of domination can be legitimated. But if political authority cannot maintain prestige in the external system, or if it cannot successfully create and deal with internal and external threats, then the greater will be its loss of legitimacy and, by the propositions in table 10.1, the more likely that conflict and change will occur.

Just as stratification is an inherently delegitimating process, at least in the long run, so geopolitics contains an inherent dialectic. This can be seen by modeling Weber's argument, as in figure 10.2.

In this model, I have taken considerable liberty to rephrase the variables in Weber's argument—always a dangerous thing to do with sociology's canonized early masters—and I have extended his ideas by the ways that I have drawn the causal arrows. What the model communicates is the reverse causal effects of external conflict which, when successful, increases prestige and legitimacy and, when unsuccessful, sets into motion the process of delegitimation. There is a built-in contradiction in the various geopolitical cycles portrayed in the model, because as the size of terri-

TABLE 10.2
Weber's Propositions on Geopolitics and Conflict

I. The greater the legitimacy of political authority, the greater its capacity to dominate other groupings in a system.
  A. The more those with power can sustain a sense of prestige and success in relations with external systems, the greater their capacity to be viewed as legitimate.
    1. The more the productive sectors of a system depend upon political authority for their viability, the more they encourage political authority to engage in military expansion to augment their interests—when successful, such expansion increases prestige.
    2. The less the productive sectors depend upon the state for their viability, the less likely they are to encourage political authority to engage in military expansion and the more likely they are to rely upon co-optation—when successful, such co-optation increases prestige.
  B. The more those with power can create a sense of threat from external forces, the greater their capacity to be viewed as legitimate.
  C. The more those with power can create a sense of threat among the majority by internal conflict with a minority, the greater their capacity to be viewed as legitimate.
II. The less a political authority can sustain a sense of legitimacy, the more vulnerable it becomes to outbreaks of internal conflict.
  A. The more a political authority loses prestige in the external system, the less able it is to remain legitimate.
    1. The less successful a political authority is in external conflict, the greater its loss of prestige.
    2. The less successful a political authority is in co-optive efforts in the external system, the greater its loss of prestige.

FIGURE 10.2. Weber's Model of Geopolitics

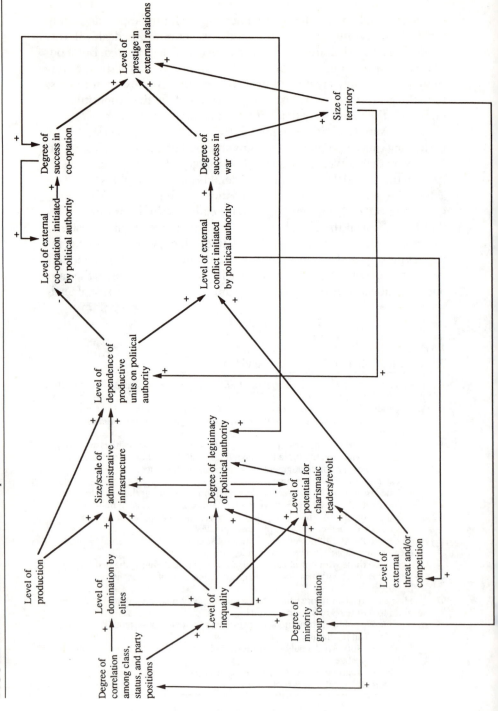

tory is extended in war, (1) the internal ethnic diversity and likelihood of internal conflict increase, and (2) the likelihood of defeat increases (through growing logistical loads, proximity of another expanding system, and hostile coalitions among neighbors). Weber does not develop this latter argument, but it is easily extracted when his ideas are modeled. Moreover, as the dependence of economic units on political authority increases with further military conquests, this positive causal loop will eventually push conflict beyond the point to which it can be successfully pursued and, hence, give prestige to political authority. In a somewhat less volatile manner, co-optation can become a self-defeating causal spiral, as success in co-optive efforts encourages additional co-optive initiatives to the point at which success becomes less likely, thereby reducing prestige and delegitimating political authority.

While Weber does not develop these ideas, except in a few pages (Weber [1922] 1968, 901–17), they add considerably to a theory of conflict, emphasizing that external system processes have important effects on internal processes, especially those revolving around integration and conflict.

CONCLUSION

Max Weber developed a suggestive and sophisticated theory of societal integration and conflict. Social integration is a function of legitimacy given to elites who dominate various social arenas, most notably in material well-being (class), power (party), and honor (status groups). The very fact of domination, especially in terms of tradition, creates stratification processes that maintain a potential for their transformation, when and if charismatic leaders can emerge and successfully mobilize subordinates. This potential is dramatically increased when legitimated political authority loses prestige in the external system, either through the unsuccessful use of coercion or through co-optation. While the existence of internal and external enemies can mitigate this situation by deflecting attention and legitimating the mobilization of coercive force, such "enemies" have to be successfully dealt with; if they are not, there is an even more dramatic loss of prestige and legitimacy. A number of more contemporary theorists have adopted these ideas—Theda Skocpol (1979) and Randall Collins (1986b, 145–84) being the most prominent—but there is obviously room for others to adopt and extend them even further. This becomes more likely, I believe, when Weber's own doubts about positivistic science are ignored and his ideas are presented as abstract models and propositions.

11 Marx and Weber Meet the Modern-day Adam Smith

Introduction

Any reasonable assessment of sociology today would conclude that the prospects for scientific theory seem rather dim. There are, of course, many theorists who advocate "positivism," or the search for general laws and models that are empirically assessed in an effort to cumulate knowledge about how the social universe operates. But if we look at sociology as an institutional whole, this is certainly not the prevailing view, especially among those who call themselves "theorists." On the research side of sociology, things are not as bad, because many advocate the use of "scientific methods" and produce suggestive findings that have theoretical implications. Yet much of the time research in sociology is atheoretical and is conducted to accommodate a patron, client, or personal interest rather than a theoretical question.

While exceptions can easily be noted, it is still reasonable to conclude that the amount of interplay between research and theory in sociology has declined. Seemingly, theory has gone one way and research another. To phrase the matter polemically, theory has increasingly been transformed into highly scholastic types of metatheory, while research has become highly routinized and ritualized often with slavish conformity to the latest research protocols, mostly quantitative in nature. Although I am overstating the point here, it would be hard to deny this trend in sociology. For we clearly are at a stage at which theory is increasingly about itself without much concern for the "real world out there,"

This chapter was originally titled "The Misuse and Use of Metatheory" and appeared in *Sociological Forum* 5 (1990): 37–53.

while research is conducted for itself without great agonizing over the cumulation of knowledge. There are specific historical reasons for this turn of events (Turner and Turner 1990), but in this chapter I want to concentrate on what this situation means for sociology in an intellectual sense. More specifically, I will examine what has happened to theory in sociology and what we should do about this sad turn of events. My target of abuse will be metatheory "as an end in itself." Metatheorizing per se does not have to be counterproductive as it is in its current practice by some sociologists who would be philosopher kings, historians, critics, commentators, or ideologues.

WHAT IS METATHEORY AND WHAT CAN IT BE?

On many occasions George Ritzer (1975, 1987, 1988, and 1991) has provided a strong defense for metatheory. Indeed, he has even offered a typology on types of metatheory (Ritzer 1988, 1991), making his efforts metametatheory. Yet over the years, I sense a change in Ritzer's position, because he seems increasingly to argue the position taken in this paper: Metatheory is best done as a means of producing better theory rather than as an end in itself. If sociology is to be a natural science, then metatheory should be a tool for generating and improving theories (Berger, Wagner, and Zelditch, 1988). Later I will illustrate this point with a protracted example, but for the present, let me ask what metatheory really is.

If we go to a dictionary and look up the word *meta*, it means "to come after"; therefore, it is reasonable to conclude that metatheory should come after we have produced some theory. Ritzer (1988) is correct to criticize Furfey (1953, 17), who sees metasociology and presumably metatheory as furnishing "methodological principles presupposed by sociology." Metatheory is not about what assumptions and presuppositions sociology should have, but about the structure and implications of existent theories. Moreover, as has been the case in physics, metatheoretical analysis should be used to generate theories that are more parsimonious, abstract, and useful in explaining how the social universe operates.

Does much metatheory in sociology seek to achieve this goal? I do not think so. Metatheory is a variety of activities with varying points of emphasis, as Ritzer (1988) points out in his typology, but my sense is that metatheory rarely involves the detailed analysis of theories in an effort to make them better. Rather, metatheory, and theory in general these days, seems to involve a variety of activities—including tracing the history of ideas, providing an intellectual biography, stating presuppositions, engaging in philosophical debate, and offering ideological critique and commentary. I am not asserting that these are *intellectually* unworthy or uninteresting activities, but I am averring that they do not advance sociology as a science. Of course, this is hardly a devastating critique for the current legions who do not think that sociology can, or should, be a science. Thus, much of what I have to say in this paper will seem irrelevant and alien to those who have a nonscientific view of sociology's agenda.

Nonetheless, let me pursue the point that the activities of many current metatheorists are scientifically counterproductive. They take us back into history of people and ideas, not social forces and dynamics; they involve us in unresolvable philosophical debates over ontologies, metaphysics, epistemology, and perhaps aesthetics; they promote an unhealthy scholasticism in which it becomes normative for theorists to produce long books with extensive quotes and footnotes that would be the envy of a biblical scholar; they offer comments on the implicit ideology and biases of this or that theory, person, or school without using such insights to produce anything but another set of ideological biases; they produce vague prescriptions that tell us such obvious things as to study action and order, conflict and consensus, structure and process, and the like; and they take theory even further away from empirical research and tests.

All these activities are, of course, intellectually interesting and stimulating, but they are not productive from the point of view of science. What, then, do I propose as an alternative? My answer to this question begins with a list of taboos that are, in essence, prescriptions for making metatheory a prelude to scientific theorizing:

1. Avoid talking about theor*ists*; instead talk about theor*ies*.
2. Avoid discussions of intellectual context, place, and time; instead, discuss social processes denoted by concepts, models, and propositions.
3. Avoid debates over philosophical issues; instead, commit one's energies to the simple assumptions that there is a world out there and that it can be understood with concepts, models, and propositions.
4. Avoid commitments to ideologies; instead, develop concepts, models, and propositions that denote operative processes in the universe (there will always be someone to expose ideological biases without your help).
5. Ignore the particulars of history; instead, examine those more general and generic processes that cut across time and place (leave something for historians to do, or if history is used, let it involve an empirical test or assessment of a theory or model).

These are simple words of advice, and in fact, they are considered naive and unsophisticated in some theory circles these days. But they are the operating assumptions of most practicing theorists in scientific disciplines. It can be wondered, of course, What is left for metatheory if these simple taboos are to be accepted by sociological theorists?

My answer is that metatheory could now (1) evaluate the clarity and adequacy of concepts, propositions, and models; (2) suggest points of similarity, convergence, or divergence with other theories; (3) pull together existent empirical (including historical) studies to assess the plausibility of a theory; (4) extract what is viewed as useful and plausible in a theory from what is considered less so; (5) synthesize a theory, or portions thereof, with other theories; (6) rewrite a theory in light of empirical or conceptual considerations; (7) formalize a theory by stating it more precisely; (8) restate a theory in better language; and (9) make deductions from a theory to facilitate empirical assessment. No one metatheorist could perform all of these activities, of course, but the point is that there would be much for various types of metatheorists to do.

To some extent, current metatheorists touch upon these activities, but not very extensively. Rather, as I noted earlier, we are far more likely to be given metatheory that emphasizes such matters as a history of a concept; an interpretation or reinterpretation of a scholar as idealist, materialist, or some such thing; a biography tracing intellectual influences on a theorist or his or her ideas; a critique of a scholar or theory as ideologically inappropriate; and so on. As interesting as such pursuits are, they can at times get in the way of clarifying the concepts, models, and propositions of existing theories, and decreasing their parsimony, elegance, inclusiveness, and explanatory power. To the extent that metatheory is concerned with these activities, I have no quarrel with it. But is this what most metatheory does? I do not think so, despite Ritzer's gallant and provocative efforts to argue the contrary (Ritzer 1988).

Let me now illustrate how metatheorizing can produce better theorizing. I do not see this example as remotely realizing every tenet listed above, but it can offer a more concrete sense of the approach that I am advocating (for longer works that come closer to realizing the ideals of these tenets, see Turner 1984, 1988).

A CURSORY AND PRELIMINARY SYNTHESIS OF THREE THEORIES

Conflict is an invariant property of social organization, because it is always present in human affairs. Conflict is also a variable, because it changes by degree and perhaps type. Thus, conflict is a proper subject for sociological theory; and, as is obvious, it has been the topic of many theories. I do not propose to develop a general theory of conflict here, but only to analyze metatheoretically three theories of conflict in order to illustrate the potential of metatheoretical analysis that is concerned with theory building.

Marx's (Marx and Engels, [1848] 1971) ideas are, of course, immediately relevant to a theory of conflict, and so we should consider his theory. Weber ([1922] 1968) is supposed to have engaged in a "lifelong dialogue" with Marx (Bendix 1968), and thus we might want to examine his conception of conflict. Finally, in his notebooks, Marx ([1857–1858] 1973) claimed that he was "improving" upon the ideas of Smith ([1776] 1937), and, therefore, we should see if there is some affinity of Marx's ideas on conflict with those of utilitarianism, especially in its modern exchange theoretic form (Homans 1961; Blau 1964; Emerson 1972; Cook 1987). What I propose is a cursory, but illustrative, metatheoretical analysis of Marx, Weber, and modern exchange theory on conflict. That is, what is the essence of these theories? Can they be stated more abstractly, generically, and parsimoniously? Can they be seen to converge or complement each other, thereby providing a more inclusive theory?

My choice of these theories is the result of my sense that, in essence, they are very similar, despite the fact that they come from very different intellectual traditions (Turner 1984, 1986). In table 11.1, I have extracted what I see as the essential concepts and propositions from the three theories. If someone does not agree with this rendering of the theories, we can compare our respective portrayals with an eye toward producing a better theory of conflict processes. Thus, part of the strategy that I advocate insists that theories be stated formally in propositions; only in this way can we have a basis for comparing them without getting bogged down in textual and scholastic debates. If portrayals of theories are couched in textual and discursive terms, as well as in the unique vocabularies of their sources, we will end up comparing unlike entities and arguing past one another—which, as far as I can tell, seems the goal of much metatheory in sociology.

Presenting these theories in table 11.1 makes evident another element of my approach: increasing the degree of abstraction to comparable levels. Marx's and Weber's theories are embedded within historical contexts—capitalist class relations and various religious movements, respectively. Exchange theory tends to be stated more abstractly, so to compare the theories, Marx's and Weber's theories must be couched at the same level of abstraction as modern exchange theories. In performing metatheory, and in trying to develop parsimonious and inclusive theory in general, the goal should always be to raise the level of abstraction. That is, we should redefine concepts in generic terms so that they pertain to all times and places rather than only to some specific historical or empirical context.

In so doing, we will often violate a theorist's intent—as is certainly the case with my rendering of Marx and Weber, who doubted that abstract and universal theory could ever be developed in the social sciences. This point should be emphasized, because much metatheory involves

TABLE 11.1
Marx, Weber, and Exchange Theory on Conflict Processes

(1) Marx's Theory of Conflict	(2) Exchange Theory of Conflict	(3) Weber's Theory of Conflict
The Conflict of Interest Principle: The greater the inequality in the distribution of material resources among actors, the greater the interest of those with resources in maintaining or increasing their share of resources and the greater the interest of those without resources in increasing their share.	*The Power-Dependence Principle:* The greater the inequality in the respective resources among social actors, the more those without resources are dependent upon those with resources in social exchanges, and hence, the greater the power of the latter over the former.	*The Discontinuity-Tension Principle:* The greater the inequality in honor-prestige, power, and material resources, and the greater the correlation across actors in the distribution of these resources, the greater the discontinuity among sets of actors and the greater the level of tension between them.
The Exploitation Principle: The greater the inequality in the distribution of material resources, the more those with a resource advantage use this advantage to extract additional resources.	*The Power-Use Principle:* The greater the power of one set of actors over another, the more that power is used to extract additional resources from the latter, and hence, to increase the degree of inequality in the distribution of resources.	
	The Justice Principle: The longer an exchange of resources endures, the more likely are norms of fair exchange to emerge and regulate the ratios of resources to be exchanged by superordinates and subordinates.	
	The Expectation-Reciprocity Principle: The longer an exchange of resources endures, the more likely are norms of reciprocity to emerge and create expectations about the ratios of resources to be exchanged in relations between superordinates and subordinates.	
The Deprivation-Immiseration Principle: The more superordinate actors use their resource advantage, the more they must (a) deny subordinates those resources necessary for their maintenance, (b) disrupt the routines of subordinates, and (c) create alienative dispositions among subordinates, and hence, the more they will increase the sense of deprivation among subordinates.	*The Deprivation Principle:* The more superordinates use their power to extract additional resources, the more they violate norms of justice, reciprocity, and expectations over rates of exchange, and hence, the more they will increase the sense of deprivation among subordinates.	*The Withdrawal of Legitimacy Principle:* The greater the correlation of resources and discontinuity between superordinates and subordinates, the more likely are the resulting tensions to lead subordinates to withdraw legitimacy from superordinates and their right to control resources.

The False Consciousness Principle: The greater the capacity of superordinates to control resources, the greater their ability to manipulate symbolic resources to mask from subordinates the sources of their sense of deprivation.

The Mobilization-Organization Principle: The more subordinates under perceived deprivation can communicate their grievances, articulate unifying symbols, and develop leadership, the more likely they are to overcome the effects of symbolic manipulation by superordinates and the more likely they are to become organized to pursue conflict.

The Violence-Redistribution Principle: The more subordinates are mobilized and organized under conditions of high inequality, exploitation, and deprivation, the more polarized are superordinates and subordinates, and hence, the greater the violence of the conflict and the greater will be the redistribution of resources among actors.

The Rational-Manipulation Principle: The more superordinates in an exchange seek to maximize their utility through power use, the more they also seek to symbolically manipulate norms of justice, reciprocity, and expectations in order to minimize the sense of deprivation among subordinates.

The Rational-Choice Principle: The more subordinates experience deprivation, the more likely they are to calculate the probability of success, the potential payoff, and the likely costs of challenges to superordinates.

The Mobilization-Solidarity Principle: The more subordinates in a system of exchange (a) can experience deprivations simultaneously, (b) can articulate unifying symbols,[1] (c) can develop leaders, and (d) can form dense and closed social networks, the more their calculations of the probable success and payoffs of challenges to superordinates will exceed their assessments of failure.

The Rebalancing Principle: The more superordinates perceive the potential success of mobilization and solidarity of subordinates, the more likely they are to renegotiate rates of exchange and to mitigate power use.

The Charisma Principle: The greater the discontinuity of resources between superordinates and subordinates, and the less the legitimacy attributed to superordinates, the more likely that charismatic leaders will emerge who can articulate deprivations, symbolically unify subordinates, and emotionally arouse subordinates to pursue conflict with superordinates.

The Routinization Principle: The greater the success of charismatic leaders in conflict with superordinates, the greater the administrative burdens of reorganizing and redistributing resources, and hence, the more formal the resulting leadership and administration of activity.

scholastic interpretations of the sacred texts—be they St. Durkheim's, St. Marx's, St. Weber's or those of some other canonized figure in our pantheon. Such interpretations are designed to show what the canonized figure "really meant," and horror is expressed about anyone who would desecrate the sacred texts. My strategy is sacrilegious, because I advocate removing ideas from their intellectual context, throwing away those that do not seem relevant or warranted for either conceptual or empirical reasons, and using only those ideas that seem to capture the dynamic of some generic process. The goal is to *use* theories to build better ones, not to become sociological monks copying and reciting passages from the sacred texts.

I have also juxtaposed the propositions of the theories in ways that emphasize their points of similarity. Those propositions that appear on the same row of table 11.1 seem to be addressing a similar dynamic, although somewhat differently. The gaps in the table show where theories diverge or examine different processes. One does not, of course, have to construct a table like this, but it does illustrate a crucial element of my metatheoretical approach: examine theories with an eye to their points of convergence and divergence, for in this way it will be easier to see how they can be synthesized and combined to produce a more powerful theory (or conversely, why they cannot be synthesized, forcing a rejection of those theoretical elements of each or all theories that seem to go astray).

Thus, Marxists, Weberians, and perhaps even exchange theorists will not like what they see in table 11.1 My goal is not to reproduce their theories but to use them for theoretical purposes. The end result should be criticized on the basis of clarity, parsimony, inclusiveness, and explanatory power, not whether a Marxian, Weberian, or exchange theoretic analysis has been produced. Most current metatheory never abandons its commitment to a scholar or theory; in my view, this unwillingness to let go of a particular theorist or a specific intellectual tradition is the great flaw of much contemporary metatheory. Eclecticism is far preferable to the current scholasticism in metatheorizing that, ironically, becomes highly parochial as scholars dare not tread outside the vocabulary or boundaries of a particular theory or intellectual tradition.

Let me now turn to the theories, as listed, but with one word of caution. Obviously, I cannot do complete justice to the theories and their constituent elements, but despite this shortcoming, the basic strategy can be illustrated. When stripped of their distracting context and complexities, each of these theories examines conflict in terms of a common set of processes: (1) inequalities in the distribution of resources; (2) differentiation of social actors into advantaged and disadvantaged positions, which in turn creates patterns of superordination and subordination; and (3) the mobilization of subordinates to initiate and engage in conflict over the distribution of resources. The theories do so in somewhat different, but nonetheless convergent and complementary, ways. In this convergence and complementarity, however, it is evident that only one subset of conflict processes is being examined—conflict in systems of inequality. Conflict is certainly a more general process, and so we should recognize that any theoretical synthesis of these three theories is only a portion of a more general theory. But in rendering a synthesis of this (and later of other manifestations of conflict), we move closer to the general "laws of conflict."

By comparing columns 1 and 2 on Marx and exchange theory, it is evident that there is very little difference between the two theories (Turner 1986). In fact, we could reasonably conclude that conflict theory is a variant of exchange theory—a metatheoretical conclusion that has many implications for some of the presumed incompatibilities among various theories. That is, utilitarian exchange theories and Marxian theories both argue that certain dynamics are inevitable in unequal systems of exchange. In fact, Marx was arguing, in essence, that the unequal systems of exchange between the bourgeoisie and the proletariat set into motion certain conflict processes. Hence, conflict theories in the Marxian tradition are exchange theories that emphasize what happens under conditions of inequality in the distribution of valued resources. Weber's theory in column 3 is far less robust than either Marx's or the exchange approach, primarily because Weber tended to view historical circumstances as highly variable and subject to chance. As a result, he felt it was not possible to specify in abstract theory the sequence of events leading to conflict; instead, he sought to provide a set of categorical concepts that could be used to interpret different empirical situations—a metatheoretical conclusion that should tell us that Weber was much less of a theorist than is typically supposed in sociological circles these days. That is, Weber tended to construct systems of categories for pigeonholing empirical/historical facts (a descriptive exercise), whereas Marx and exchange theorists developed abstract propositions and principles to explain *why* these facts reveal a certain character (a theoretical exercise). Such was not always the case with Weber, however, as is evident in table 11.1. For here Weber does address conflict in theoretical terms, and the skeletal theory plugs into that of Marx and exchange theory at crucial places, signaling that Weber clearly understood the most important dynamic forces behind conflict in systems of inequality.

Where the theories differ is particularly interesting, because here they can perhaps provide a corrective for each other. For example, Marx's predictions about violence and proletarian revolutions go astray, I suspect, because he failed to give adequate weight to the rational choice principle and the rebalancing principle. He assumed actors would incur

FIGURE 11.1.   Toward a Synthesis of Conflict Theories in Systems of Inequality

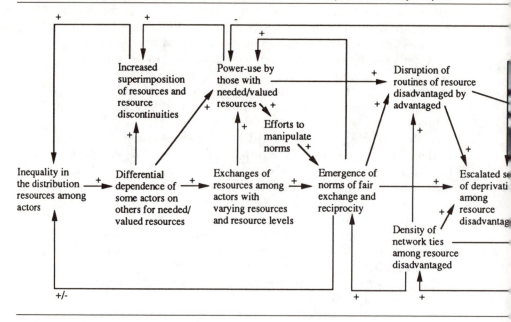

the costs of revolt and that superordinates in capitalist systems could not mitigate power-use and make compromises. He also mistakenly assumed that organization always leads to violence, whereas in fact organization and articulation of goals sometimes lead to compromise (violence comes with emotional arousal and incipient organization, emerging but unsettled leadership, and emotionally laden but imprecise ideological articulation; see Simmel [1903–1904] and Coser [1956] on this point). At other times, of course, a sense for obvious empirical evidence informs us that organization can lead to violence, as is the case of civil wars and wars between societies. But *within a system of inequality*, organization will tend to produce compromise if it exists before the onset of violence. However, if violent conflict is initiated and persists before the conflict parties have become organized (as is the case with many civil wars), then it will tend to increase the organization of the conflict parties *and* the level of violence.

In addition to providing potential avenues of correction for each other, a comparison of theories allows for synthesis into an even more robust theory about conflict between superordinate and subordinate actors in systems of inequality. There are many paths to synthesis, but the one that I prefer is to construct analytical models in which the variables articulated in theories are arrayed in visual space as a complex configura-

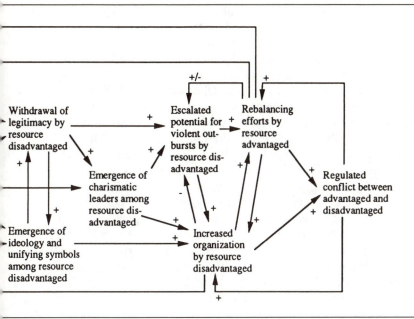

tion of causal processes (see Turner 1984 and 1985a, for a more complete statement on analytical models). In turn, the key causal connections in the model can serve as a stimulus to reformulating the propositions. One could, of course, move directly to a synthesis at a propositional level, but the use of an analytical model provides an additional tool for mapping the processes by which, or through which, the variables in a proposition are connected. Moreover, a model is more parsimonious, because it is possible to simultaneously present configurations of relations (whereas propositions, especially verbal ones, are more awkward when they seek to do this; mathematical statements, however, can achieve greater parsimony in stating multiple relationships among variables).

At any rate, figure 11.1 presents my best effort at combining and extending the three theories presented in table 11.1. The terms in the boxes should be viewed as variables; and in reading them, one can insert the words *degree of (power use, superimposition, inequality,* etc.). I have simply phrased the variables in an active way to emphasize the causal processes connecting these variables. The sign for each causal arrow (+, –, + / –) denotes positive (+), negative (–), or positive or negative relations depending on certain conditions (+ / –). This notation system is admittedly awkward and would be refined in a more detailed analysis.

Moreover, the relations among these variables are more complex than indicated in the model, but this complexity is more readily captured in propositions. When I view it as a process, I see three important configurations, roughly corresponding to the left, middle, and right clusters of variables in the model presented in figure 11.1. Let me review each of these, starting with the causal processes on the left of figure 11.1.

As all exchange theories argue, inequalities among actors in their control of valued resources create patterns of dependence that, in turn, give those who control valued resources power over those who do not. Such power is typically used to extract additional resources and to symbolically manipulate the perceptions of those who are dependent. Such power use (Emerson 1972) will increase the correlation (or superimposition) among those holding valued resources as well as the degree of discontinuity between advantaged and disadvantaged resource holders (Weber's variables). If no other forces were operating, inequalities would constantly escalate—the "rich getting richer and the poor poorer." To some extent, this set of causal processes exerts enormous pressure in systems of inequality to escalate resource differentials, but there are also powerful countervailing forces operating to mitigate this pressure. One such force is the emergence of norms about fair exchange (often conceptualized as "justice" in exchange theories) and reciprocity (or the requirement to provide, in turn, an expected level of resources even to those in a disadvantaged bargaining position). That is, unabated power use (Marx's exploitation), even under conditions of symbolic manipulation (Marx's false consciousness), will eventually violate norms of justice and reciprocity, setting into motion the processes in the middle of figure 11.1 (this is another reason why Marx's predictions go astray). Before examining this portion of the model, I should stress the negative feedback arrows from these normative processes to power use, because these normative forces often work by themselves to check power use. At times, the advantaged have internalized norms, and hence impose limits on their exploitive action on moral grounds. At other times, they simply calculate how far they can go, or the potential consequences of excessive power use (e.g., conflict with or, if possible, withdrawal by subordinates). Thus, a system of inequality can achieve a short-term equilibrium if inequalities correspond to norms of fair exchange and reciprocity, and if those in advantaged positions resist fully exploiting their advantage or can control the production of symbols so as to redefine the nature of fairness and reciprocity. This latter is more difficult to implement, and it often backfires when the disadvantaged compare the old and newly proposed rates of exchange from a highly moralistic view. When this occurs, norms of fair exchange and reciprocity may accentuate a sense of deprivation among those in disadvantaged positions.

Moving further into the middle portions of figure 11.1, power use, per se, can escalate subordinates' sense of deprivation, because as subordinates are asked to give more, their routines are disrupted. And if such disruption is compared against the yardstick of norms of fair exchange and reciprocity, the resulting sense of deprivation is heightened even more. The density of network ties (especially rates of interaction and lines of communication) is a critical variable in these processes. Marx conceptualized this process in terms of propinquity of the proletariat in urban areas, whereas more analytical versions of Marx (e.g., Dahrendorf 1959) conceptualize this process in terms similar to those in figure 11.1. The essential points are these: (1) Dense networks will produce and sustain (through informal social control) normative expectations about what is fair, just, normal/routine, and reciprocal; (2) dense networks will also be highly conducive for the communication of grievances, thereby escalating the sense of deprivation among those connected in the network; and (3) both directly and indirectly through its effects on other variables, network density is structurally conducive to the emergence of ideologies and other unifying symbols, to the withdrawal of legitimacy and support from the existing system of resource exchange, and to the emergence of charismatic leaders who can articulate grievances, foster ideological codification, and encourage the withdrawal of legitimacy. These latter variables obviously come from Marx and Weber in their efforts to explain the mobilization of subordinates to pursue conflict. In exchange terms, actors must feel sufficiently deprived, uncommitted, and justified (1) to perceive that their current situation is below what is minimally acceptable (the Thibaut and Kelley [1959] notion of "comparison level" is relevant here) and (2) to believe the costs of organized protest are less than the potential rewards (the redistribution of resources). Such calculations are most likely when actors can communicate and resonate their grievances in dense networks, formulate unifying symbols, and develop spokespersons who can articulate their grievances and offer alternatives to the present situation.

Turning to the right portion of the figure, it is at this point that violence, physical coercion, or mass outbursts are most likely. Contrary to Marx's position, however, the organization of subordinates into a coherent structure, with clear leaders, agenda, and goals, will reduce this potential for violence. If subordinates are not well organized, but experience escalations in their sense of deprivation, especially as measured against norms of fairness as these have been ideologically loaded by leaders and by interaction in dense networks, the resulting withdrawal of legitimacy can lead to sudden outbursts of mass, collective, and violent protest. Specific incidents can set this potential off into overt action, but often superordinates will, instead, make concessions, reduce power use, and thereby forestall collective outbursts. Moreover, if sporadic and episodic violence occurs, such violence is frequently followed by increased organization, bargaining, and negotiating. The result is that concessions are made on both sides, and the conflict becomes regulated or institutionalized. That revolutions, at the societal level, are so rare would signal that the potential for violence, or its localized and episodic emergence, sets into motion less volatile forces: rebalancing by the advantaged faced with an increasingly organized and unified set of subordinate actors, or a willingness to compromise on the part of subordinates as their goals and the costs of achieving them become more clearly articulated (what Coser [1967, 37–52] has called "realistic" conflict). As is evident, however, feedback processes are decisive in the regulation or deregulation (violence) of the conflict. Rebalancing is a complicated process, and, as the feedback loops indicate with the plus/minus sign, it can often escalate relative deprivation (the plus side); and at other times, it can lower this sense of deprivation (the minus side) if the withdrawal of legitimacy and ideological arousal have not gone so far as to strip elites of credibility and good faith. But as conflicts become regulated, or as subordinates become more organized, such organization can increase network density, which, in turn, can lower deprivations since there is now an institutionalized means for redressing grievances (hence, the plus/minus sign from the "density" to "deprivation" variable). Also, as power use declines as a result of rebalancing, or as network densities have organization superimposed on them thereby creating new norms of fair exchange and reciprocity that actually legitimate existing inequalities, the system of exchange portrayed on the left of figure 11.1 can be reequilibrated, at least for a time.

CONCLUSION

Such is the general line of argument suggested by Marx's and Weber's theories and exchange theory, as the ideas in these theories have been subject to metatheorizing that produces theory, as opposed to "discourse." Even if my rendering of these theories invites argument, or "discourse," it can focus on specific variables, their interconnection, and their isomorphism with actual empirical events. In my view, such is the proper "use" of metatheory *when the goal is to produce scientific sociology*. It would be a "misuse" of metatheory to criticize me for desecrating Marx, Weber, and one's favorite exchange theorist, or to insist that I cite this or that source, or to require that I elaborate on philosophical issues. These kinds of attacks are a misuse of metatheory, because they will not produce theory; they will, instead, produce philosophy, history of ideas, debate, and commentary. All these activities may be intellectually interesting, but they *do not* increase our understanding of the operative dynamics of the universe. And here is where I part company with defenders of metatheory like Ritzer, because in the end I would argue that sociology must be a science and that other activities in sociology—including metatheory itself—must be subordinate to this goal. Many sociologists want sociology to be "anything and everything for everyone," whereas I want it to be a science. And I believe most current metatheory does not contribute to this end.

Thus, this rendering of the three theories is intended to promote science, as opposed to scholarship, which can border on scholasticism in some sociological circles. Science involves developing abstract propositions and models; my approach to metatheory is (1) to construct propositions and models and (2) to move back and forth between the two. In this case, I began with propositions and converted them into a model. With a model like that in figure 11.1, it is possible to convert the critical causal paths back into propositions. Such propositions can select the most central paths and, at the same time, specify with more precision the nature of the relations (for example, linear, curvilinear, logarithmic, or exponential) among the variables (see Turner 1984 for a further explanation, or Collins 1988, 157–58, for a secondary summary). For example, I think that three of the most crucial "laws" that emerge from the synthesis in figure 11.1 are the following:

1. *The Law of Conflict Potential.* The degree of conflict potential in a system of inequality is a multiplicative function of (a) the degree of resource superimposition and discontinuity, (b) the level of power use by those in resource-advantaged positions, and (c) the degree of salience and efficacy of norms of fair exchange and reciprocity.

2. *The Law of Conflict.* The level of overt conflict between the resource advantaged and the resource disadvantaged in a system of inequality is a multiplicative function of (a) the extent to which power use by superordinates disrupts the routines of subordinates and violates norms of fair exchange and reciprocity; and (b) the degree to which such power use escalates the sense of deprivation among subordinates.

3. *The Law of Conflict Violence.* The level of violence in the conflict between the resource advantaged and the resource disadvantaged in a system of inequality is (a) an exponential function of the degree of power use by superordinates and a logarithmic function of escalation in the deprivations of superordinates and their withdrawal of legitimacy from the existing system of inequality; and (b) a curvilinear function of the degree of clarity, leadership, codification of ideology, and formality in organization among subordinates.

We might wish to propose other laws—say, for instance, one on conflict institutionalization or rebalancing by superordinates. I view these processes as subsumed by the three laws stated above, but others might disagree. But the critical point is that any disagreement is over theory—that is, models and propositions stating basic relations among forces in the universe. The goal of metatheory is to debate at this level and to avoid the long-winded and debilitating "discourse" that typifies metatheory today and, sad to say, most social theory in general.

The above review is certainly not the final word on Marx, Weber, and exchange theory, nor is it a very detailed or precise metatheoretical analysis. The goal in this chapter has been to provide not a full-blown execution of my strategy, but rather a modest illustration of one way in which metatheory can produce theory. Others are invited to improve upon the theory presented above by doing a better job of metatheorizing.

12 Spencer, Marx, Weber, Simmel, and Pareto on Power and Conflict

The insights of such scholars as Karl Marx, Émile Durkheim, and Max Weber on social theory in general and political sociology in particular are well documented. Other early masters of sociological theory, such as Georg Simmel, Herbert Spencer, and Vilfredo Pareto, are given somewhat less attention in both theory and political sociology. Yet all of these thinkers developed a number of converging and complementary "laws" of political organization. And despite the widespread attention given to some of these historical figures, few attempts have been made at extracting these laws.[1] One reason for this oversight is that some, such as Marx and Weber, did not believe that abstract and universal laws on invariant properties of the social universe could be developed. This belief is shared by many sociologists. Another reason is that the ideas of various scholars are seen as "intellectual totalities," and it is considered inappropriate, if not sacrilegious, to extract only portions of a scholar's thought in an effort to isolate what is considered theoretically useful.

There are, no doubt, many other reasons for the unwillingness to examine Marx, Weber, Spencer, Durkheim, Pareto, and Simmel as social theorists who articulated some of sociology's first laws of social organization. Our purpose here is not to delineate the reasons, but to view these scholars as theorists and to extract from the corpus of their work those abstract principles and laws that still seem useful in political sociology. Naturally, Marx and Weber would turn over in their graves if they knew

This chapter is coauthored with Charles H. Powers and originally appeared under the title "Some Elementary Principles of Political Organization: Insights from Sociology's First Masters" in *Research in Political Sociology* 2 (1986): 1–17.

that this kind of exercise was performed on their work; and a good many living sociologists and political scientists are similarly unsympathetic to such exercises.

Yet, to the extent that sociology is considered a science, and to the degree that the term *science* in *political science* is taken seriously, we should try to extract the theoretical principles from scholars' work, state them formally, and use them in our theoretical and research efforts to build a cumulative science. While many do not feel that emulating "the natural sciences" is possible or appropriate, we believe that creating a "natural science of society" is a reasonable project. Accordingly, in this essay, we will try to demonstrate the utility of our view by articulating some elementary principles of political organization that can be found in the works of selected scholars—Herbert Spencer (1876, 447–588), Karl Marx and Friedrich Engels ([1848] 1955), Karl Marx ([1867] 1967), Max Weber ([1920] 1968), Émile Durkheim ([1893] 1933), Georg Simmel ([1907] 1978, [1908], 1956), and Vilfredo Pareto ([1901] 1968, [1909] 1971, [1916] 1935, [1921] 1984). When principles are extracted and abstracted, the surface incompatibility of various scholars' work is dramatically reduced. Indeed, there is an amazing degree of convergence in the ways these classical scholars viewed political processes. We should caution, however, that these principles will seem familiar because they have been used in much contemporary theory and research. Often, contemporary practitioners have themselves implicitly extracted these principles, but equally often—and tragically, from our viewpoint—we have had to rediscover these principles, because of an unwillingness to perform the present exercise earlier and more frequently.

Naturally, we can make only a modest beginning in using the past masters to develop political theory. Accordingly, we will confine our discussion to what we perceive to be the most interesting principles with respect to (1) social system differentiation, (2) political mobilization, (3) political oscillation, and (4) political conflict. All these principles reveal some of the conditions that influence the form of political organization in social systems. And while these principles are from sixty to a hundred years old, we will indicate some of the ways that they inform contemporary theorizing in sociology and political science.

PRINCIPLES OF SOCIAL SYSTEM DIFFERENTIATION

Early sociological theorists clearly recognized that the degree and nature of political organization in a social system are related to its level of structural complexity. The more that roles and social units are differentiated, the greater the problems of their coordination and control, but also the greater the potential for the use of power. Thus, one of the basic questions that early thinkers such as Spencer, Marx, Durkheim, Simmel, Pareto, and Weber all addressed is: What conditions increase the level of differentiation in social systems? Their answers varied somewhat, but they can be summarized as principle I:[2]

I. The degree of differentiation in a social system is a positive and additive function of

A. the level of productivity in the system (Marx [1867]; Pareto [1901] 1968, [1909] 1971), with productivity being a positive and additive function of

1. the level of technological knowledge (Marx [1867]; Simmel 1890),
2. the level of access to material resources (Marx [1867] 1967),
3. the degree of secularism and acquisitiveness emphasized in cultural symbols (Pareto [1916] 1935; Weber [1922] 1968);

B. the degree of competition for valued resources, with competition being a positive and additive function of

1. the absolute size of the population (Spencer 1876; Durkheim [1893]),
2. the degree of ecological concentration of a population (Spencer 1876; Durkheim [1893]),
3. the degree of perceived scarcity in material and symbolic resources (Durkheim [1893]),
4. the relative size of elite segments of a population (Pareto [1909] 1971, [1916] 1935);

C. the degree to which previously differentiated social units have become integrated, with integration being a positive and additive function of

1. the degree of functional interdependence among social units (Spencer 1876; Durkheim [1893]),
2. the degree of consensus over cultural symbols (Durkheim [1893] 1960; Weber [1922] 1968),
3. the availability of symbolic media of exchange (Simmel [1907] 1978),
4. the capacity to mobilize and culturally legitimate political power (Marx and Engels [1848] 1955; Spencer 1876; Durkheim [1893]; Pareto [1916]; Weber [1922]).

These principles link differentiation, economic productivity, competition for resources, and previous patterns of social integration. That is, high levels of productivity, competition, and integration are the generic conditions under which high levels of social differentiation are to be found. In turn, each of these three conditions is related to other forces. Productivity is dependent upon levels of technology and resource availability. Competition is related to a population's size and concentration, to its sense of resource scarcity (money, land, natural resources, power, prestige, honor, etc.), and to the number of elites who appropriate resources and thus are viewed as competitors. Success at integrating previous patterns of social differentiation is a result of developing common values and beliefs, generalized media of social exchange (such as money and contracts), and the capacity to mobilize power to regulate transactions among differentiated social units.

While these principles may seem "obvious," they are nonetheless some of sociology's basic laws of the social universe. Moreover, they provide a starting point for understanding political processes in social systems. First, these principles connect the degree of political organization to the level of differentiation in social systems. Second, the *form* of political organization is related to those variables—such as productivity, competition, and integration—that reveal high or low values. For example, political organization will reveal one form (i.e., unstable and dictatorial) when productivity and consensus over values and beliefs are low while competition for resources is high, whereas another form (i.e., stable and democratic) will emerge when there is high productivity, low competition, and high value consensus. Moreover, changes in the respective values of the variables will alter what we can expect to occur in social systems. In our view, one of the tasks of political theorists is to specify the conditions that influence the values of the variables in these principles as they relate to increases (or decreases) in the mobilization and use of power.

This principle of structural differentiation in social systems has, we feel, implications for much of the current literature in political sociology. For example, there exists a large literature devoted to "institution building," especially on the problems and dilemmas of the Third World (McHenry 1979; Mazrui and Tidy 1984), but most of this work remains uninformed by abstract theoretical principles. There is an equally large literature on the structural conditions favoring (or discouraging) political democratization, and yet, much of this diverse literature (Almond and Verba 1965; Moore 1966; Tilton 1975) does not appear to have drawn great insight from early sociological theory.

A brief and cursory review of some hypotheses in this literature can, we feel, illustrate our point. For example, one hypothesis is that "social and economic development" is a prerequisite for the spread of "political democracy" (Cutright and Wiley 1969–1970; Dahl 1971). This large literature focuses on the importance of factors such as literacy, mass communication, and involvement in the commercial sector. Increased modernization is thought to produce conditions favoring organizational complexity, democratic order, and political stability. A second hypothesis concerns "political culture." The development of complex and democratic political institutions is in some respects contingent upon the values, sentiments, and orientations of the common citizen. Political development is least likely when people are most parochial and passive (Banfield 1958; Almond and Verba 1965). A third prominent hypothesis is that rapid social change destabilizes sociopolitical institutions (Sofranko and Bealer 1972).

Although a number of interesting case studies are now being generated in an effort to sort out these hypotheses (Putnam et al. 1983), there does not appear to be an overarching theoretical framework that makes sense of empirical findings and meaningfully interrelates competing hypotheses (Huntington and Dominguez 1975). We feel that sociology's early masters provided some guidelines for making these more general theoretical statements. For example, to translate some of the key concepts in principle I, diffusion of technology, discovery of resources, and accumulation of material and human capital are typically seen to spur economic growth. Economic growth and complexity create environmental conditions necessitating political development (Powers 1985). Population density and scarcity promote competition, which in turn undermines parochial ways of thinking and forces peasants to develop new ways of looking at the world (Farb 1978). Modernization can be destabilizing but does not have to be. Organizational interdependence reduces instability (Sofranko and Bealer 1972), as does the production of unifying symbols (Schwartz 1982). Thus, many of the main findings and competing hypotheses emerging from the literature on institution building in political sociology can be interpreted in terms of principle I, based on the writings of Marx, Engels, Pareto, Simmel, Weber, Spencer, and Durkheim. Indeed, we feel that this principle represents at an abstract level the more generic variables and relationships that have been proposed by a variety of political theorists. In essence, principle I provides a *structural* interpretation of "institutional development" by linking social development (and by implication, political development) to levels of social differentiation, which, in turn, are connected to the conditions listed in IA, IB, and IC. The connection between political development and the structural basis of "institution building" becomes even more evident when we turn to principle II dealing with political mobilization.

PRINCIPLES OF POLITICAL MOBILIZATION

As we have indicated above, early sociologists saw political organization as inhering in social structural differentiation (the structural process behind institution building). Social differentiation inevitably results in increased power that is potentially available for mobilization and in organizational forms capable of mobilizing that power. At the same time, the concentration of power is influenced by other system dynamics, such as the mobilization of counterpower and the generation of social conflict. Early sociologists distinguished between various *forms* of social differentiation in an effort to understand the ways in which the organization of power inheres in the overall structure of society. That is, depending not only on the degree but also on the form of social differentiation, political organization will vary.

Probably the most critical variables influencing the nature of political mobilization were best conceptualized by Spencer (1876), Marx and Engels ([1848] 1955, [1867] 1967), and Durkheim ([1893] 1933). Each classical sociologist emphasized a different variable, thereby rendering their respective schemes one-sided. Yet, taken together, their ideas can be translated into principle II:

II. The degree of development in the regulatory structures that mobilize and use power is a positive and additive function of
   A. the number, volume, and rate of internal activities among system units, with these being a positive and additive function of
      1. the absolute number of differentiated units (Spencer 1876; Durkheim 1893),
      2. the level of productive activity in each unit (Marx [1867] 1967; Spencer 1876; Pareto [1916], 1935),
      3. the level of distribution of both information and materials among units (Spencer 1876);
   B. the degree of external threat perceived to exist in the environment of a system (Spencer 1876; Simmel 1903–1904, 1908);
   C. the degree of internal threat perceived to exist within the system, with perceptions of internal threat being a positive and additive function of
      1. the degree of dissimilarity in the goals of system units (Spencer 1876),
      2. the degree of rank differentiation (Mark and Engels 1848);
   D. the lack of consensus over abstract cultural symbols (Durkheim 1893).

In this principle, the degree of political organization is seen as connected to the productive and distributive activity among diverse units in the system. The more the activity, the greater is the need to regulate and coordinate these activities. Perceived threats (Spencer 1876; Simmel 1908) also encourage the elaboration of political organization, since it takes the consolidation of power and its implementation through organizations (such as military and domestic bureaucracies) to reduce actual or imagined threats. Finally, there is a limit to how far political organization can be carried without supporting and legitimating cultural values, beliefs, and ideologies. Marx and Engels ([1848] 1955), Spencer (1876), Pareto ([1916] 1935, [1921] 1984), Weber ([1922] 1968), and Durkheim ([1893] 1960) all emphasized the view that political organization inevitably generates tension, resistance, and countermobilization of power, which often erupts into conflict between those with and those without resources. To mitigate this inevitability, consensus over cultural symbols becomes increasingly necessary as a condition for further political mobilization (Durkheim [1893] 1933).

Just how political mobilization occurs in a specific empirical case is, of course, beyond the capacity of abstract theory to document. But the principles articulated above provide the general theorems from which deductions to particular cases can be made. Depending on the values of these variables, different degrees and forms of political mobilization are likely. It is toward specifying the theoretical consequences of different weightings of the variables in principle II that both theory and research in political science and sociology should be directed. For example, as Simmel (1908) and Spencer (1876) emphasized, highly centralized political mobilization is likely under conditions of external threat. Or, to take another example, coercive and centralized systems are likely when consensus over cultural symbols is low, especially under conditions of internal threat. Thus, the abstract theorems provided by the first masters may, at first glance, seem obvious and trivial, but on the contrary, they provide insight into the relations among the generic variables from which further deductions to specific cases can be made.

In fact, most of the existing literature on political mobilization is an empirical description of the processes outlined at a more abstract level in principle II. This literature tends to focus on the political instability of societies during the transition from agrarian to industrial forms of development (Moore 1966; Migdal 1974; Tilly 1978). That is, much of the political mobilization literature concentrates on the IIC portion of principle II, whereas the first masters emphasize other processes (IA, IIB) as equally important. Thus, the insights of these early sociologists might be seen as providing additional variables for research on "political mobilization" in differentiating (principle I) social systems.

PRINCIPLES OF POLITICAL OSCILLATION

One of the most frequently rendered observations on political processes is that political organization tends to oscillate between centralized and decentralized profiles. Actually, the long-run trend appears to be toward ever greater centralization of power at the societal level of social organization, with periodic efforts to decentralize power in order to forestall for a time further centralization.

Some of the early theorists sought to understand the dialectical forces inherent in both centralized and decentralized forms of political organization. For scholars such as Spencer (1876) and Pareto ([1921] 1984), centralized power generates pressures for decentralization, whereas decentralized forms of power activate processes encouraging centralization. Unfortunately, much of the commentary on Spencer and Pareto has retained their awkward vocabulary (such as "lions," "foxes," "speculators," "retiers," "militant," and "industrial") and has not extracted the more generic properties of power that these concepts denote.

The basic insight of Spencer, Pareto, and, to a lesser extent, Durkheim is that decentralized political systems create integrative problems of coordination and control. These problems can stem from the lack of unifying cultural symbols, the diversity of productive and distributive activity, and the overuse of co-optation[3] as a means of political regulation. These dynamics are stated below in principles III, IV, and V:

III. The greater the level of decentralization of political power in a social system, the greater
   A. the diversity and inconsistency of cultural symbols (Durkheim 1893; Pareto 1916);
   B. the diversity of productive and distributive activity (Spencer 1876; Pareto 1909, 1916);
   C. the use of co-optation as a means of political regulation (Pareto 1901, 1916).
IV. The greater (a) the diversity of productive and distributive activity, (b) the level of diversity in cultural symbols, and (c) the use of co-optation as a means of political regulation in a social system, the greater the problems of coordination, control, and integration in that system (Pareto [1901] 1968, [1916] 1935, [1921] 1984).
 V. The greater the problems of coordination, control, and integration in a social system, the more likely is the mobilization and consolidation of power in that system (Spencer 1876; Pareto [1916] 1935).

Conversely, political centralization sets into motion pressures for decentralization. Centralized power involves direct regulation, often backed by the use (or threat) of coercion. Moreover, it tends to restrict the variety of productive and distributive activities to conform to political directives, and it seeks to articulate and impose conservative cultural symbols. Over time, these restrictions create tensions and pressures for decentralization, which will often be resisted (thereby creating further centralization of power), but which will eventually result in some decentralization. If such decentralization does not occur, more intense forms of conflict can result (see next section). These insights are formalized in principles VI, VII, and VIII:

VI. The greater the level of centralization of political power in a social system, the greater
   A. the use of direct regulation and (the threat of) coercion (Spencer 1876; Pareto 1901, 1916, 1921);
   B. the restrictions on productive and distributive activities (Spencer 1876; Pareto 1909, 1916);
   C. the efforts to articulate and impose conservative cultural symbols (Durkheim [1893], 1933; Pareto [1916], 1935).
VII. The greater the level of (a) direct regulation and/or the use of coercion, (b) constraint on productive and distributive activities, and (c) imposition of conservative cultural symbols, the greater the level of resentment against the agents of centralized political power (Spencer 1876; Pareto 1901, 1916, 1921).
VIII. The greater the level of resentment against the agents of centralized political power, the greater the pressures for decentralization of political power (Spencer 1876; Pareto 1901, 1916, 1921).

Political oscillations between centralized and decentralized system profiles are thus the result of certain processes that inhere in one of these two states. We have phrased the propositions in a way that avoids previous interpretations of Spencer and Pareto. These interpretations inappropriately emphasize the inevitability of the full cycle from a centralized to a decentralized system profile, and vice versa. (While Pareto's work connotes this inevitability, Spencer's does not; and even Pareto was more cautious than is often recognized.) Much more important for our purposes is their recognition of the inherent pressures for centralization in decentralized systems and for decentralization of centralized systems. Just whether this cycle is completed depends upon the empirical values of the variables in a particular system. But Spencer and Pareto clearly recognized the critical relationships among the use of power, the diversity of productive and distributive activity, the nature of cultural symbols, the levels of social integration, and the degree of accumulated resentment among members of a population.

These principles, we feel, provide the foundation for much of the current research literature on political sociology. For example, there is considerable controversy over such issues as "the inevitable rise of oligarchy" (Michels [1915] 1959), the existence of "power elites" (Hunter 1953;

Mills 1959; Rose 1967; Dahl 1971; Useem 1984), the centralization of "monopoly capital and power" (Baran and Sweezey 1966; Szymanski 1981), the expansion of the "bureaucratic state and its control systems" (Hage and Aiken 1967; Collins 1975), and similar topics concerning the way power becomes concentrated in social systems (Dye 1983). Principles III through VIII place many of these debates in the research literature into a theoretical context. What they do is emphasize the dialectical nature of power—that is, centralization produces pressures for decentralization, and vice versa. Moreover, they specify the conditions that produce these pressures for decentralized systems (principle IV) and centralized systems (principle VII). Thus, depending upon the point in the cycle when cross-sectional research is undertaken, we suspect that different results will obtain. And rather than viewing highly centralized or decentralized political systems as inevitable, these principles point to systematic fluctuations in the *degree of* concentration in power. As such, they provide interesting leads for research that can avoid many of the problems of overgeneralization in the research literature (see Klapp 1975).

In particular, the principles direct attention to the diversity in (a) cultural symbols, (b) productive activities, and (c) co-optation as variables that influence the direction of the centralization–decentralization cycle (principles II and IV). And so, rather than end states in themselves, these points in the cycle appear to be conditions that produce tendencies toward their opposite. Surprisingly, for all of the emphasis on "the dialectic of power" at the conceptual level in social theory, research has tended to ignore the view that centralization–decentralization are processes. We think that these principles of political oscillations, therefore, can reorient the research literature on the topic of how and when power becomes concentrated.

PRINCIPLES OF POLITICAL AND SOCIAL CONFLICT

All theorists of the nineteenth century recognized that the concentration of power activates processes for the mobilization of counterpower. This mobilization is often translated into decentralization of power as delineated in principles III through VIII above. But under certain conditions, the concentration of power generates active resistance that escalates into open conflict between the agents of political authority and opposition groups. The combined theoretical legacy of Marx, Simmel, and Weber gives us some clues as to what these conditions are.

The key conflict-producing dynamics reside in the respective degrees of concentration in, and the strength of the correlation among, political, material, and symbolic resources. The greater the nonrandom concentrations and distribution of these resources, and the higher the correlation among them, the greater the potential for conflict (Marx and Engels 1848, Weber 1922). Just what conditions are considered critical in translating such inequality into conflict vary from theorist to theorist, but a composite formulation of their ideas can be expressed in the following principles:

IX. The degree of inequality in a social system is a positive and joint function of
   A. the level of concentration of political, material, and symbolic resources (Marx and Engels 1848);
   B. the strength of the correlation in the distribution of political, material, and symbolic resources (Weber 1922).

X. The degree of incompatibility of interests among segments in a population is a positive function of the degree of inequality (Marx and Engels 1848).

XI. The degree to which incompatibilities in interests become translated into conflict is a positive and additive function of
   A. the level of awareness among the deprived and subordinate segments of a population about the extent of inequalities, with such awareness being positively and additively related to
      1. the level of alienation in subordinate populations (Marx and Engels 1848),
      2. the degree of disruptive change in the daily routines of subordinate populations (Marx and Engels 1848),
      3. the degree of communication among subordinate populations (Marx and Engels 1848),
      4. the extent of restrictions on upward social mobility (Marx and Engels 1848, Weber 1922),
   B. the availability of charismatic leaders to (a) codify resentments of subordinate populations into a unified ideology, and (b) arouse the emotions of subordinates (Marx and Engels 1848, Simmel 1908, Weber 1922);
   C. the capacity of subordinates to become politically organized, with such organization being positively related to (a) and (b) above and negatively related to
      1. the degree of organization among superordinate populations (Marx and Engels 1848).
      2. the degree of consensus among subordinates and superordinates over abstract cultural symbols (Pareto [1916] 1935, Weber 1922).

XII. The degree of violence in the conflict between superordinates and subordinates in a social system is positively related to XIA and XIB above (Marx and Engels 1848; Simmel 1908) and negatively related to XIC above (Simmel 1908).

Principles IX through XII summarize at an abstract level some of the basic arguments of Marx, Weber, and Simmel on the topic of conflict. According to these first masters, conflict ultimately emanates from inequalities and the capacity of subordinates to become aware of their situation, to develop leaders who can mobilize ideological sentiments, and to organize on behalf of their cause. However, contrary to Marx's (Marx and Engels [1848] 1955) assertions, the degree of violence of such conflict is negatively related to political organization among subordinates. Violence is most likely to occur, according to Marx, when subordinates are aware of inequalities, become emotionally aroused, and develop political organization. With organization, however, clear leaders, goals, and programs are outlined, with the result that superordinates and subordinates are more likely to bargain and compromise. Yet, once violence is initiated, it can force subordinates to organize and pursue further violence against the superordinate's agents of coercion.

The research literature on conflict processes, of course, is enormous (Coser, 1967; Shelling 1971; Kriesberg 1982). It ranges from the study of societal revolutions (Davies 1962; Gurr 1970; Paige 1975; Kelley and Klein 1977) to the mobilizations for change-producing social movements (McCarthy and Zald 1977; Jenkins 1983). In many respects, these principles on conflict processes from the early masters have already been incorporated into the research literature. The revolutionary conflict literature tends to emphasize "relative deprivation" and other conditions listed in principle XIA, whereas the social movements literature stresses the conditions in XIB and XIC. Thus, it may well be the case that the principles of these early masters will continue to offer guidance for the growing research literature on conflict in political sociology. But it is nonetheless important, we feel, to articulate the principles clearly so that we can appreciate which portions have become well integrated into research literature.

CONCLUSION

In this chapter, we have not sought to present a comprehensive theory. Rather, we have articulated in rough form twelve basic theoretical principles from the early masters on how and why political organization increases, oscillates, and generates conflict. Most of these principles had been clearly articulated before the turn of the last century, and they stand at the core of our understanding about political processes in social systems. Of course, they require supplementation, but our intent has been to extract only the essence of Spencer's, Marx's, Simmel's, Durkheim's, Pareto's, and Weber's thoughts on political structure and dynamics. Obviously, these great scholars addressed additional issues and developed principles on other properties of social systems.

We have briefly commented on the relevance of these principles to certain research traditions in political sociology, but we have not tried to summarize the entire literature. Our sense is that only in the area of political conflict have the principles of the early masters been fully incorporated into, and extended by, the research literature (for a thorough coverage of the recent political sociology literature, see Weil and Dobratz 1984). The other principles, we feel, provide some interesting leads for researchers. In particular, we see it as desirable to incorporate more explicitly the process of structural differentiation (principle I) into theoretical and research activity on political mobilization and oscillation (principles II through VIII).

Of course, it should not surprise us that these principles require supplementation and reformulation. They are, after all, rather old. But if only to highlight what these first masters gave us as a theoretical legacy, the exercise in this chapter, we believe, has been worthwhile. The real work now begins: to reformulate principles I through VIII, as has been implicitly done for those on conflict (principles IX through XII), in ways that revitalize them and make them useful in research on political processes.

NOTES

1. An exception is the author's efforts in J. Turner (1978) and Turner and Beeghley (1981) to develop theoretical principles.

2. This and the other eleven principles presented in this essay reflect a certain vision of theory building. See J. Turner (1978, 1980) for a more complete discussion. It should be emphasized that these are macroprinciples and therefore address only the most basic affinities among generic properties of social systems. They are the theorems of a formal system of theoretical deduction. They also set the parameters within which microanalysis can be profitably pursued.

3. For our usage here, co-optation is the method of gaining voluntary compliance and cooperation from differentiated units by making sure that the vested interests of those units are interdependent.

13 George Herbert Mead's Behavioral Theory of Social Structure

George Herbert Mead's sociological work (1934, 1938) is typically viewed as providing the conceptual underpinnings of modern symbolic interactionism (Blumer 1962, 1969). It is also seen as contributing to other theoretical perspectives, including role theory (R. Turner 1968, 1978, 1979), dramaturgy (Goffman 1959), and even phenomenology (Schutz 1971). In adopting Mead's ideas, all these perspectives have, it is argued in this chapter, underemphasized two facets of Mead's work: (1) his behaviorism, which, as Baldwin (1986) has only recently pointed out, is highly contemporary; and (2) his structuralism, which, as Nisbet (1974, 108–27) has argued virtually alone, is compatible with the work of such figures as Durkheim (1893).

The underrepresentation of Mead's behaviorism and views on social structure has been unfortunate, not only because of the resulting distortion of Mead's scheme, but also because of its role in perpetuating the dualism between principles of human behavior and those of social structure. Indeed, much contemporary psychological and sociological theory consists of gallant assertions about the primacy of behavior (Homans 1974), action (Parsons 1937), symbolic interaction (Blumer 1962), and macrostructure (Blau 1977). Although such assertions have allowed scholars to concentrate their theoretical efforts and to generate interest-

This chapter was originally titled "A Note on George Herbert Mead's Behavioral Theory of Social Structure" and appeared in the *Journal for the Theory of Social Behaviour* 12 (1982): 213–22.

ing insights, they have also tended to polarize conceptual thinking in ways that inhibit understanding of (1) the behavioral basis of social organization; and (2) the organizational basis of behavior. Mead's ideas provided one way to reconcile the conceptual schism between structural and behavioral approaches in sociology.

G. H. MEAD'S BEHAVIORISM
The Basic "Life Process"

Mead's behaviorism represents a synthesis of several convergent philosophical traditions: Darwinism, pragmatism, and utilitarianism (Turner and Beeghley 1981, 464). From Darwin, Mead adopts the idea that social life represents a continual process of adjustment and adaptation to the environment; from pragmatism, especially John Dewey's (1922) instrumentalism, Mead emphasizes the process of thought and reflection as humans adjust to their environments; and from utilitarianism, Mead stresses that behavior is goal directed and involves mental calculations of utility, pleasure, and pain. Thus, the basic "life process," as Mead phrased the matter, involves goal-directed adjustments to environments in terms of thought, reflection, and calculations of rewards. Much human behavior is, therefore, covert, leading Mead to devote considerable energy to rejecting John B. Watson's extreme position (Mead 1934, 10):

> Watson apparently assumes that to deny the existence of mind or consciousness as a psychical [sic] stuff, substance, or entity is to deny its existence altogether, and that a naturalistic or behavioristic account of it as such is out of the question. But, on the contrary, we may deny its existence as a psychical entity without denying its existence in some other sense at all; and if we then conceive of it functionally, as a natural rather than transcendantal phenomenon, it becomes possible to deal with it in behavioristic terms.

For Mead, then, the unique "subjective capacities" of humans are behaviors and can be understood in terms of reinforcement principles. In Mead's view, the most rewarding states for humans are adjustment and adaptation to social environments; and hence, the behavioral capacities of humans arise and are retained by virtue of the reward value of adjustment to social structures.

The Basic Behavioral Capacities

The behavioral capacities for using "conventional" or "significant" gestures, for "role-taking," for "mind," and for "self" are the result of reinforcement in social contexts. That is, these behaviors arise from, and are sustained by, the reward value of adjustment and adaptation to social structures (Mead 1934). And as will be examined shortly, it is these capacities that create and sustain ongoing patterns of social organization. It is his analysis of these behaviors and their consequences for human organization that makes Mead's principles a "behavioristic theory of social structure."

"Conventional" or "Significant" Gestures.

In contrast to Darwin's view that gestures only signal emotions, Mead follows Wundt's insight that gestures mark and punctuate the course of all ongoing activity (Mead 1934, 42–68). In fact, action and interaction involve a "conversation of gestures" in what he termed "the triadic matrix" that reveals the following three phases (Mead 1934, 81): (1) gestural emission of one organism as it acts on its environment, (2) a response of another organism to these gestures of the acting organism, which, in turn, becomes a gestural stimulus to the acting organism, and (3) an adjusted response by the acting organism as it takes into account the stimuli of the responding organism. Both "lower" and "higher" organisms engage in this conversation of gestures; and it is in this conversation that "meaning" resides, since meaning is simply the way in which organisms are prepared to act on their environment. That is, meaning is "not to be conceived, fundamentally, as a state of consciousness, or as a set of organized relations existing or subsisting mentally outside the field of experience into which they enter; on the contrary, it should be conceived objectively, as having its existence entirely within the field itself" (Mead 1934, 78).

The difference between human and nonhuman use of gestures, Mead argues, is that humans use "conventional" or "significant" gestures which evoke the same "meaning" or tendency to behave in both the acting and the responding organisms. This fact gives the triadic matrix, or "the conversation of gestures," an added dimension, since organisms can use gestures to evoke the same responses in each other. This capacity allows them to coordinate their responses; in Mead's eye, it is the behavioristic basis for social organization. In contrast, nonhuman animals, Mead emphasizes (and probably overemphasizes),[1] simply respond to each other's gestures; that is, the gestures do not "mean the same thing" (that is, activate the same behavioral response) to the acting and responding organisms. As he noted, the "roar of the lion does not mean the same thing to it and its potential victim," whereas if someone shouts "Fire!" in a crowded place, this gesture evokes the same responses in the sending and receiving organisms.

Role-taking.

The capacity to use conventional gestures provides the behavioral impetus for the development of another critical capacity: "role-taking," or "taking the role of the other" (Mead 1934). This capacity is the most fundamental concept in both Mead's behaviorism and his view of social structure. By reading each other's significant gestures, humans can assume each other's perspective, or "attitudes," and hence each other's disposition to act, with the result that they can mutually adjust their lines of conduct and cooperate.

Mind.

The use of significant gestures and role-taking are, in Mead's view, adjustments and adaptations to organized social contexts. They are retained behaviors, because they are rewarding in that they facilitate adaptation to others in groups. Once the use of significant gestures and role-taking become part of the behavioral repertoire of an organism, then minded behavior, or "mind" in Mead's terms, becomes increasingly likely. In turn, as minded behavior expands, the use of conventional gestures and role-taking also become more extensive, complex, and elaborate. For Mead, as for Dewey (1922), "mind" is not an entity but a process of behavioral adaptation to the environment. It involves the capacity to engage in covert behavior and to "imaginatively rehearse" alternative lines of conduct, to visualize their consequences, and to select responses that facilitate adjustment to the environment (Mead 1934, 42–125). Mind is not inborn or an entity; it is a behavioral response that expands with biological maturation, the use of conventional gestures, role-taking, and efforts to adjust to social contexts. It is a learned response to an environment of conventional gestures among others organized into groups; and as the facility for minded behavior expands, it allows for adjustment to even more diverse and complex social contexts.

Self.

Like mind, self is a behavior. Moreover, it is a behavioral capacity that is possible only with organisms that have some facility at minded behavior. Borrowing from Cooley and James, Mead visualizes self as the ability of organisms to see themselves as objects in their environment. This behavior emerges as humans role-take with others, sense the attitudes of others toward them, assess themselves in relation to these attitudes, and evaluate themselves in accordance with these attitudes. By seeing themselves as objects in an environment of others prepared to act in certain ways, human achieve greater facility for adjusting their responses to one another. The capacity to read the gestures of others, to perceive oneself as an object in relation to these gestures, and then to enter this perception into the "imaginative rehearsal" process (mind) increases the ability to calibrate and fine-tune overt responses to others.

Over time, Mead visualizes that a "complete" or "unified" self emerges. As people role-take, see themselves as objects, and adjust their responses, they come to visualize themselves as certain types of objects, and they develop stable "meanings" about (or dispositions to act toward) themselves as objects. Such responses further facilitate adjustment, because they give people's actions a typical style and form, as well as a high degree of predictability. The ability to predict another's typical pattern of response greatly increases cooperation and mutual adjustment.

The Behavioral Basis of Social Organization

Mead conceptualized social organization, or "society" in his terms, as stable patterns of behavior and interaction among individuals. As such, society constitutes the environment in which humans learn—through the reinforcement that comes with adjustment and cooperation—the behavioral responses involved in the use of significant gestures, role-taking, mind, and self. Yet, once this behavioral repertoire exists, it constitutes the underlying behavioral basis for society. For while the unique abilities of humans can emerge only from necessary adaptation to society, social organization could not be sustained without these behaviors.

Most discussions of Mead's concepts end here, without specifying in any detail how the structural properties of social organizations depend upon variations in the behavioral capacities for the use of conventional gestures, role-raking, mind, and self. It is the failure to draw out the social structural implications in Mead's writing that has, it is argued in this chapter, created an unnecessary conceptual gulf between micro- and macro-analysis in sociology.

G. H. MEAD'S BEHAVIORAL STRUCTURALISM

For Mead, the basic structural properties of "society" are "scope" and "complexity." While his discussion of these topics is unfortunately brief (Mead 1934, 317–36), Mead does provide a rough sketch of his ideas on the behavioral basis underlying increases in the scale of social organization. These are presented in his discussion of the "obstacles and promises in the development of the ideal society," but even in this more philosophical or moralistic context, a set of behavioral principles on the conditions increasing or decreasing the scale, scope, and integration of social systems emerges from his discussion. By "scale" and "scope," Mead emphasizes size and differentiation; and by "integration" Mead denotes "cooperation" and "regularized interaction" among diversely situated individuals. Hence, his principles concern specifying those conditions that increase the level of cooperation among diverse individuals in systems of varying size and complexity.

For Mead, cooperation among individuals increases when

1. they can use a common set of significant symbols in social contexts and thereby invoke common responses in each other;
2. they can accurately role-take with each other and thereby determine each other's dispositions to act in social contexts;
3. they can covertly rehearse alternative lines of conduct, inhibit inappropriate responses, and emit responses that will facilitate their mutual adaptation to a social context (mind); and
4. they can use the "attitudes of others," as well as a more stable self-conception of oneself as a certain types of object, as a source of self-evaluation and self-control in social contexts (self).

As is evident, these conditions of cooperation follow from the behavioral capacities of humans for using conventional gestures, for role-taking, for mind, and for self. The scale of society and its degree of integration are thus a result of the degree to which conditions 1–4 above are met. That is, the size of a population and the level of social differentiation among individuals in that population are limited by the degree to which variations in conditions 1–4 can integrate individuals. If people cannot use common conventional gestures, role-take accurately, rehearse alternatives, and see themselves as stable objects of evaluation in situations, then social control and integration are likely to be difficult to maintain. Yet, as society increases in size and level of differentiation, the maintenance of common gestures, the accuracy of role-taking with diverse others, the rehearsal of alternatives in different and often unfamiliar situations, and the ability to see oneself as a consistent object in changing contexts all become problematic and pose a severe integrative dilemma.[2]

How, then, is the dilemma resolved? Mead's answer sounds very Durkheimian (1893) at first glance; but upon closer inspection, it is more sophisticated than Durkheim's, because it specifies the behavioral processes underlying integration in larger and more complex systems. The critical concept in Mead's view is what he terms "the generalized other" or "community of attitudes" among collectivities of individuals (Mead 1934, 152–63). This concept has inspired the development of specific theoretical ideas, such as reference group theory (Shibutani 1955), but its implications for a more general theory of social structural integration have not been fully drawn out. The reason for this oversight is that Mead's discussion of the generalized other initially occurs in his discussion of the "stages" in the development of self (Mead 1934, 152–64) and is only invoked again near the very end of *Mind, Self, and Society*. For Mead, individuals must be capable of more than interaction in "play" and "game" situations; they must also be able to role-take with the "generalized other" and use this "community of attitudes" for self-evaluation and for self-control:

> The complex co-operative processes and activities and institutional functionings of organized human society are also possible only insofar as every individual involved in them or belonging to that society can take the general attitudes of all other such individuals with reference to these processes and activities and institutional functionings, and to the organized social whole experiential relations and interactions thereby constituted—and can direct his own behavior accordingly. (Mead 1934, 155)

While Mead's ideas may not seem to go beyond Durkheim's insistence that social integration is possible only to the extent that individuals are guided by a "common conscience" or "collective conscience" (Durkheim [1893] 1933), Mead's scheme specifies the underlying behavioral processes by which people can be guided by a "common conscience" or "generalized other." That is, by reading each other's gestures and by role-taking, humans assume not just the role of specific others in a social context but also the general perspective of the collective enterprise; and then they use this "generalized other" as a major force in minded deliberations and in self-evaluations. In this way, people not only adjust their conduct to each other in an immediate context, but also constrain these person-to-person adjustments in terms of a broader perspective or "community of attitudes." Thus, in contrast to Durkheim's *ex cathedra* pronouncement that the collective conscience is "external" and "constraining" on individuals (Durkheim [1895] 1938), Mead provides a behavioral basis for understanding the nature of such constraint.

With this basic argument, Mead examines the variations in the scale and integration of social systems (Mead 1934, 317–36). Mead hypothesizes that large-scale social systems composed of many individuals playing diverse roles will develop numerous subgroups, each of which will reveal its own, somewhat distinctive, "generalized other." Hence, any complex society will evidence multiple generalized others, which individuals use as a framework in adjusting their conduct to each other. Yet it is possible for such multiple generalized others to be contradictory, resulting in a situation in which individuals can potentially come into conflict. Thus, the scale of society is limited by the degree to which multiple generalized others do not bring individuals, who role-take with these "others," into conflict. How is this avoided? For Mead, several processes operate to mitigate potential conflicts, and in so doing, promote integration.

One process is the development of common conventional gestures among individuals who are diversely located in different groups (Mead

1934, 321–22). With common symbols, individuals can invoke similar responses in one another and role-take more accurately. In Mead's view, to the extent that diversely situated individuals (even those whose interests are in conflict) can use similar gestures, they can place themselves in each other's perspective; and through the capacities for mind and self, they can adjust their responses to each other (Mead 1934, 326–27). Conversely, to the extent that members of diverse groups do not share a repertoire of significant gestures, then role-taking will be difficult, and societal integration will be problematic.

Another process facilitating the integration of larger and more differentiated systems is the development of a highly abstract generalized other that encompasses the more specific generalized others of different subgroups (Mead 1934, 327–28).[3] In this way, individuals in otherwise different situations and contexts can role-take with a common generalized other and use this very abstract perspective to guide their mutual adjustments. But if an abstract perspective that cuts across the more specific perspectives of diverse subgroups cannot be developed and used as a basis for role-taking, minded deliberations, self-evaluation, and self-control, then social integration becomes problematic.

Thus, from Mead's analysis, the scale and integration of the social order are connected to the development of common conventional symbols among diversely located individuals and the utilization of both specific and abstract generalized others as a frame of reference. To the degree that subgroups cannot develop their own generalized other, role-taking, minded deliberations, and self-evaluations become problematic and inhibit people's cooperation. To the extent that common symbols are not used, role-taking among differentiated populations becomes problematic and thereby increases the potential for antagonistic relations. And to the degree that an abstract generalized other that cuts across subgroups and their more specific frameworks cannot be developed, then role-taking once again becomes problematic, because individuals in different groups do not have common community attitudes to use in adjusting their responses to each other. But unlike Durkheim's (1893, 1904) similar pronouncements on social integration and its various pathologies, Mead's ideas are tied to a behavioral theory, and thus, Mead can be labeled a "behavioral structuralist."

CONCLUSION

The goal of this chapter has been to redirect attention to two facets of Mead's thinking, namely behaviorism and structuralism. Much of Mead's behavioristic emphasis has been ignored by social psychologists, and his analysis of society is rarely cited in more structural approaches.

Structural sociology rarely explores the behavioristic basis of emergent phenomena, whereas behaviorists have taken several decades to enter the "black box" and seem ill-disposed to venture in the other direction toward an analysis of social structures. Yet Mead saw little polarity between structure and behavior. For Mead, behavioral capacities emerge from the reinforcement achieved with adjustments to social structures, whereas the operation of social structures depends upon these acquired behavioral capacities. Mead is thus one of the few social theorists ever to posit a behavioral theory of social structure.

By way of summarizing the previous discussion, Mead's theory can be presented in more formal terms:

> The degree of organization (cooperation) among pluralities of individuals and the scale of their organization are positive and additive functions of their capacity
>
> 1. to employ a common repertoire of significant gestures in all interactive contexts
> 2. to utilize significant gestures to role-take simultaneously with
>   a. varieties of specific others who are present in an immediate setting
>   b. varieties of generalized others that reveal different degrees of generality
> 3. to imaginatively rehearse alternative lines of conduct, inhibit inappropriate responses, and select for emission those responses that will facilitate cooperation
> 4. to see themselves as objects in situations and to use evaluations of themselves in terms of the attitudes of specific and generalized others as a basis for regulating responses

Thus, as Mead ventured into the "black box" that Watson eschewed, he also provided a way to examine not only subjective processes but also social structural processes. While Mead's ideas are not well developed, they provide a conceptual lead for reconciling what are often viewed as antagonistic approaches to building social theory. This chapter has sought to present Mead's scheme with the hope that behavioral and structural social theorists can utilize Mead's general framework for further speculation.

NOTES

1. Mead probably viewed humans as overly unique. Studies of primates and other higher mammals reveal that they also possess, in less developed form, the behavioral capacities that Mead saw as exclusively hu-

man. For Mead argued that "in man the functional differentiation through language gives an entirely different principle of organization which produces not only a different type of individual but also a different society" (Mead 1934, 244).

2. This dilemma is phrased in a manner similar to Adam Smith's original formulation in *The Wealth of Nations*—a formulation that was repeated by French structuralists like Saint-Simon, Comte, and Durkheim.

3. Durkheim (1893) makes the same argument in his discussion of how the collective conscience grows more "abstract" and "general" as systems assume an "organic" basis (differentiated) of solidarity.

CHAPTER

14 Mead as a Social Physicist

Unlike practitioners of the more mature sciences, sociologists do not agree on the subject matter of their discipline, and they hold widely varying opinions on how theory should be constructed and what it is supposed to accomplish. This eclecticism is not, I feel, a healthy sign; rather, it is a clear indication that sociology is in a state of intellectual confusion. Sociologists have lost their vision of what science is. Indeed, only in a discipline that has lost its way could mechanical number crunching, per se, be considered "science" and philosophical navel contemplation be defined as "theory."

It is almost as if we have forgotten that science and theory are part of the same enterprise. That is, science is to seek understanding of the universe, and the vehicle through which such understanding is to be achieved is theory. Sociology has allowed poor philosophers to usurp theoretical activity and "statistical packages" to hold social science hostage. In this chapter, I would like to communicate an alternative to this bifurcated and polarized situation. This alternative is not new; on the contrary, it is what Auguste Comte (1854) proposed in 1831 for "the science of society." In proposing this alternative, I will discuss some of the problems in what passes for sociological theory. And then, I will illustrate my alternative by converting the work of an acknowledged philosopher, George Herbert Mead, into sociological theory. I have chosen Mead to illustrate my argument because his work has been enormously influential and because he is of a generation that was unafraid to ask the basic questions of all theoretical activity: What is the nature of the universe? and what concepts and principles will allow us to understand this universe?

This chapter was originally titled "Returning to Social Physics: Illustrations from the Work of George Herbert Mead" and appeared in *Current Perspectives in Social Theory* 2 (1981): 187–208.

Whether one is a physicist or a sociologist, these must be our guiding questions. For while we may seek to understand different dimensions of the universe, our goal is the same: to develop principles that allow us to understand *why* the universe operates in certain regular ways.

GEORGE HERBERT MEAD AS A "SOCIAL PHYSICIST"

Mead was, of course, a philosopher, not a self-conscious sociologist. Yet he is still read today because he uncovered some of the most fundamental processes in the social universe. He was, I feel, the epitome of what sociologist theorists should be. He borrowed concepts, without great concern for "moods of the time" or the "intellectual milieu" of those thinkers whose work he found most useful. He articulated a clear philosophical position—a mixture of behaviorism and pragmatism—that became translated into a clear theoretical argument and that can be translated into explicit principles. He was concerned not with causality, but with the fundamental relations among critical processes in the social universe.

His concepts present problems of operationalization; they do not yield precise predictions; and they do not try to explain every aspect of the universe. Rather, Mead sought to articulate the relationship among (1) human behavioral capacities for mind and self; (2) human action; (3) human interaction; and (4) human social organization. He was, in my view, a "social physicist" who saw that understanding of the social world can be achieved only by developing concepts and propositions that capture the key relationships among generic properties of social organization.

Most treatments of Mead fail to recognize the explicitly theoretical aspects of his work. I would like to draw attention to Mead's theoretical principles by, first of all, summarizing briefly the assumptions from which these principles follow. Then I will articulate, as a series of abstract principles, Mead's insights into the nature of human capacities for mind, self, action, interaction, and social organization. "Meadians" and others may find this treatment distasteful, but it is my feeling that this kind of scrutiny of Mead is overdue. And the reasons for many scholars' distaste for what I will do is at the heart of what is wrong with sociology as a theoretical discipline. We worship people, not concepts and principles. We fall into the morass of metaphysics, rather than using assumptions as a way to isolate concepts and principles.

Mead's General Assumptions

Mead was a behaviorist, but one who criticized the extremes of John B. Watson and other extreme behaviorists. Like Watson, he rejected the notion that "mind" and other covert behaviors are "things," but he refused to be drawn into Watson's efforts at excommunicating covert behaviors from reality. As Mead emphasized (1934, 10):

> Watson apparently assumes that to deny the existence of mind or consciousness as a psychical stuff, substance, or entity is to deny its existence altogether, and that a naturalistic or behavioristic account of it as such is out of the question. But, on the contrary, we may deny its existence as a physical entity without denying its existence in some other sense at all; and if we then conceive of it functionally, and as a natural rather than transcendental phenomenon, it becomes possible to deal with it in behavioristic terms.

For Mead, "mind" and "self" are *behaviors* that develop in humans as a result of their neurological capacities and as a consequence of their adjustment to patterns of social organization. But once developed, it is these behavioral capacities that enable humans to act and interact in ways that sustain patterns of social organization. Thus, Mead's more purely sociological work revolves around an analysis of humans' unique behavioral capacities for mind and self and how these make human action, interaction, and organization possible.

Mead's Principles of Action

For Mead, all organisms, whether human or nonhuman, act on their environment in an effort to meet their needs. In *The Philosophy of the Act* (1938), Mead presented several general principles on what is involved in initiating and giving direction to behavior in *all* organisms. His views are expressed as "stages" by the editors of *The Philosophy of the Act*, but Mead's ideas are better expressed as a series of principles (Mead 1938):

I. The greater the degree of maladjustment of an organism to its environment, the stronger are its impulses.

II. The greater the intensity of an impulse for an organism, the greater is the organism's perceptual awareness of objects that can potentially consummate the impulse and the greater is its manipulation of objects in the environment.

    A. The more maladjustment stems from unconsummated organic needs, the greater the intensity of the impulse.

    B. The longer an impulse goes unconsummated, the greater the intensity of the impulse.

III. The more impulses have been consummated by the perception and manipulation of certain classes of objects in the environment, the more likely are perceptual and behavioral responses to be directed at these and similar objects when similar impulses arise.

These three principles summarize Mead's behaviorism. Action emerges out of adjustment problems encountered by an organism. Such problems were termed "impulses" by Mead. Behavior is directed at restoring equilibrium between the organism and the environment. The essence of behavior involves perception and manipulation of objects. And successful perception and manipulations are retained in the behavioral repertoire of organisms. As Mead argued, it is from this behavioral base that human action emerges, but the capacities for mind, self, and society require additional theoretical principles if the distinctive qualities of human action are to be understood. But these additional principles are best outlined after Mead's general principles of interaction among organisms without mind, self, and society are delineated. For as with action, human interaction is an extension of more general processes of interaction among organisms without mind and self.

Mead's view of the fundamental process of interaction for *all* organisms, whether human or nonhuman, can be expressed in two additional principles:

IV. The more organisms seek to manipulate objects in their environment in an effort to consummate impulses, the greater the visibility of gestures that they emit during the course of their action (Mead 1934, 42–46).

V. The greater the number and visibility of gestures emitted by acting organisms, the more likely are organisms to respond to each other's gestures and to adjust responses to each other (Mead 1934, 42–46).

These two propositions underscore Mead's view that the essence of interaction involves (a) an organism emitting gestures as its act on the world, (b) another organism responding to these gestures and hence emitting its own gestures as it seeks to consummate its impulses, and (c) readjustments of responses by each organism on the basis of the gestures emitted. Mead termed the process the "triadic matrix," or "conversation of gestures," and it can occur without cognitive manipulations and without the development of common meanings. Indeed, as Mead argued, it is only among humans with mind and self, living in society, that this fundamental interactive process involves cognitive manipulations, normative regulation, and shared meanings.

Principles of Human Action, Interaction, and Organization

Critical to understanding Mead's view of human action, interaction, and organization is the recognition that humans develop mind and self out of their participation in society. Thus, Mead's formulation of principles on the development of mind and self will precede a discussion of how these two behavioral capacities alter the nature of action, interaction, and organization for humans.

Principles on the Emergence of Mind.

Any particular individual is born into a society of actors with mind and self (1934, 133). In attempting to understand how infants come to adjust to adult actors and to society, Mead offered a series of important principles on the process of socialization. The first of these principles deals with the emergence of mind:

I. The more an infant must adapt to an environment composed of organized collectivities of actors, the more likely the infant is to be exposed to significant gestures (1934, 133).

II. The more an infant must seek to consummate its impulses in an organized social collectivity, the more likely is the learning of how to read and use significant gestures to have selective value for consummating the infant's impulses (1934, 68–70).

III. The more an infant can come to use and read significant gestures, the greater its ability to role-take with others in its environment, and hence, the greater its capacity to communicate its needs and to anticipate the responses of others on whom it is dependent for the consummation of impulses (1934, 76–82).

IV. The greater the capacity of an infant to role-take and use significant gestures, the greater its capacity to communicate with itself (1934, 61–67).

V. The greater the capacity of an infant to communicate with itself, the greater its ability to covertly designate objects in its environment, inhibit inappropriate responses, and select a response that will consummate its impulses and thereby facilitate its adjustment (1934, 90–91).

VI. The greater the ability of an infant to reveal such minded behavior, the greater its ability to control its responses, and hence, to cooperate with others in ongoing and organized collectivities (1934, 91).

These propositions should be read in two ways. First, each proposition, by itself, expresses a fundamental relationship in the nature of human development. For example, proposition I states that human infants are, by virtue of being born into society, inevitably exposed to a collage of significant gestures; and proposition II states that since infants must consummate impulses in a world of significant gestures, they will learn to read and use these gestures as a means of increasing their adjustment. Thus, each proposition states that one variable condition, stated in the first clause of the proposition, will lead to the development of another capacity (stated in the second clause of the proposition) in the maturing human infant. Second, the sequence of six propositions should be viewed as marking "stages" in the genesis of a critical behavioral capacity, mind. For Mead, mind emerges out of a connected series of fundamental processes summarized in the six propositions above.

Principles on the Emergence of Self.

As the capacities for mind begin to emerge, the capacity for self also becomes evident (1934, 134–35). However, the full development of self is, like the development of mind, the result of a series of fundamental processes. These processes are summarized by the following propositions:

VII. The more a young actor can engage in minded behavior, the more it can read significant gestures, role-take, and communicate with itself (1934, 133).

VIII. The more a young actor can read significant gestures, role-take, and communicate with itself, the more it can see itself as an object in any given situation (1934, 136–37).

IX. The more diverse the specific others with whom a young actor can come to role-take, the more it can increasingly come to see itself as an object in relation to the dispositions of multiple others (1934, 158–59).

X. The more generalized the perspective of others with whom a young infant can come to role-take, the more it can increasingly come to see itself as an object in relation to general values, beliefs, and norms of increasingly larger collectives (1934, 138).

XI. The greater stability in a young actor's images of itself as an object in relation to both specific others and generalized perspectives, the more reflexive its role-taking and the more consistent its behavioral responses (1934, 140).

  A. The more the first self-images derived from role-taking with others have been consistent and noncontradictory, the greater the stability, over time, of an actor's self-conception (1934, 144).

  B. The more self-images derived from role-taking with generalized perspectives are consistent and noncontradictory, the greater the stability, over time, of an actor's self-conception (1934, 144).

XII. The more a young actor can reveal stability in its responses to itself as an object, and the more it can see itself as an object in relation to specific others as well as generalized perspectives, the greater its capacity to control its responses, and hence, to cooperate with others in ongoing and organized collectivities (1934, 156–58).

These propositions document Mead's view of certain fundamental relationships among role-taking acuity, self-images of oneself as an object, and the capacity for social control. They also summarize Mead's conceptualization of the sequence of events involved in generating what he termed a "unified self" (1934, 144) in which an individual adjusts its responses in relation to (a) a stable self-conception, (b) specific expectations of others, and (c) general values, beliefs, and norms.

As the consecutive numbering of the propositions underscores, the development of mind and self is a continuous process. And once mind and self in individual human organisms have developed, the nature of action and interaction, as well as patterns of social organization, among humans is qualitatively different from that of nonhuman organisms. Yet Mead emphasized that there is nothing mysterious or mystical about this qualitative difference. Indeed, even though human action, interaction, and organization are distinctive by virtue of the capacity for mind, self, and symbolically mediated organization into society, this distinctiveness has been built upon a base common to all acting organisms. The elaboration of this base necessitates, of course, new principles to describe the unique nature of human action, interaction, and organization. And it is in the elaboration of these additional principles that Mead provided a number of critical insights into how and why humans create, maintain, and change patterns of social organization.

Principles of Human Action and Interaction.

The emergence of mind and self complicates somewhat Mead's view of the act as involving impulse, perception, and manipulation, as well as his notion of the triadic matrix as a simple process of organisms emitting and reading gestures as they adjust their responses to each other. Indeed, the complications introduced into the processes of action and interaction make for an entirely new way to organize a species.

In regard to action, we can combine Mead's insights and articulate one additional principle to account for the distinctive features of human acts.

I. The greater the intensity of impulses of humans with mind and self, (a) the more likely is perceptual awareness of objects that can potentially consummate the impulse to be selective; (b) the more likely is manipulation to be covert; and (c) the more likely are both perception and manipulation to be circumscribed by a self-conception, expectations of specific others, and generalized perspectives of organized collectivities.

When action, as described in proposition I above, occurs in a social context with others, then it becomes overt *inter*action. But it should be emphasized that even isolated acts, in which others are not physically present, involve interaction with symbolically invoked others and generalized perspectives. The capacities for mind and self, Mead argued, assure that humans will invoke the dispositions of others and broader "communities of attitudes" to guide behavior during the course of their acts, even if specific others are not physically present and even if others do not directly react to one's behaviors. But when others are present, the use of significant symbols and role-taking becomes more direct and immediate, requiring several supplementary propositions on interaction (Mead 1934, 133–34, 222–26):

  II. The more humans with mind and self seek to consummate impulses in the presence of others, the more likely they are to emit overt significant gestures and the more likely they are to read the significant gestures of others, and hence, the greater their role-taking activity.

  III. The more humans role-take with each other, the more likely is the course of their interaction to be guided by the specific disposition of others present in a situation, by the self-images of oneself as a certain type of object in the situation, and by the generalized perspective of the organized collective of others in which they are participating.

These three principles summarize the fundamental relationships that Mead saw as inhering in the triadic matrix of human interaction (Mead, 1934–69). When stated as principles, the key relationships among impulses, significant gestures, role-taking, self-conceptions, expectations of others, and generalized perspectives are highlighted.

Principles of Human Social Organization.

Since interaction among humans is possible by virtue of role-taking abilities, and since society involves stabilized patterns of interaction, society for Mead is ultimately a process of (1) role-taking with various "others" and (2) using the dispositions and perspectives of these others for self-evaluation and self-control. The nature and scope of society, Mead implicitly argued, are a dual function of the number of specific others and the abstractness of the "generalized others," or what I termed "generalized perspectives," with whom and with which individuals can role-take. In many ways, Mead viewed society as a "capacity" for various types of role-taking. If actors can role-take with only one other at a time, the capacity for society is limited, but once they can role-take with multiple others, and then with generalized others, their capacity for society is greatly extended. These fundamental relationships are summarized in Mead's two basic propositions on the dynamics underlying society:

I. The more actors can role-take with pluralities of others and use the dispositions of multiple others as a source of self-evaluation and self-control, the greater their capacity to create and maintain patterns of social organization (Mead 1934, 263–65).

II. The more actors can role-take with the generalized perspective of organized collectivities and use this perspective as a source of self-evaluation and self-control, the greater their capacity to create and maintain patterns of social organization (Mead 1934, 264).

If actors cannot meet the conditions specified in these two "laws" of social organization, then instability and change in patterns of interaction are likely. Actors who cannot role-take with multiple others at a time, and use the dispositions of these others to see themselves as objects and to control their responses, will not be able to coordinate their responses as well as actors who can perform such role-taking. And actors who cannot role-take with the general norms, beliefs, values, and other symbol systems of organized groups, and use these to view themselves and to regulate their actions, will not be able to extend patterns of social organization beyond immediate face to face contact. For it is only after actors can role-take with what Mead termed a broader "community of attitudes," and use a common set of expectations to guide their conduct, that extended and indirect patterns of social organization become possible.

In addition to these two basic principles, Mead elaborated several propositions on role-taking with generalized others. Since the scope of society is ultimately a positive function of role-taking abilities with generalized others, Mead apparently felt it necessary to specify some of the variables influencing the relations among role-taking, generalized others, and the nature of society. Three variables are most prominent in Mead's scheme: (1) the degree to which actors can hold a *common* generalized other (Mead 1934, 265–68), (2) the degree of *consistency* among multiple generalized others (Mead 1934, 322), and (3) the degree of *integration* among different types and layers of generalized others (Mead 1934, 322–27). These variables were implicitly incorporated by Mead into additional principles of social organization.

III. The more actors can role-take with a common and generalized perspective and use this common perspective as a source of self-evaluation and self-control, the greater their capacity to create and maintain cohesive patterns of social organization (Mead 1934, 321–22).

In this principle, the ability to role-take with a common perspective (norms, values, beliefs, and other symbolic components) is linked to the degree of cohesiveness in patterns of social organization. Thus, unified and cohesive patterns of organization are maintained, Mead argued, by a common collective perspective. On this score, Mead came close to Durkheim's emphasis on the need for a "common conscience" or "collective conscience." But in contrast to Durkheim, Mead was able to tie this point of emphasis to a theory of human action and interaction, and hence he was in a position to specify the mechanisms by which individual conduct is regulated by a "generalized other" or a "collective conscience."

Much like Durkheim, Mead also recognized that the size of a population and its differentiation into roles influence the degree of commonality of the generalized other. These variables can be expressed as two sub-propositions (Mead 1934, 321–30):

IIIA. The more similar the position of actors, the more likely are they to be able to role-take with a common and generalized perspective.
IIIB. The smaller the size of a population of actors, the more likely are they to be able to role-take with a common and generalized perspective.

Naturally, the converse of propositions III, IIIA, and IIIB could signal difficulties in achieving unified and cohesive patterns of social organization. If the members of a population cannot role-take with a common generalized other, then cohesive social organization will be more problematic. And if a population is large and/or highly differentiated, role-taking with a common generalized other will be more difficult.

Like Durkheim (1893, 1897), Mead recognized that a common "generalized other" becomes increasingly tenuous with growing size and differentiation of a population. For large differentiated populations there are multiple "generalized others," since people participate in many different organized collectivities. These considerations led Mead to view consistency of "generalized others" as related to how extensive differentiation of social structure could become.

IV. The more actors can role-take with multiple but consistent generalized perspectives, and the more they can use these perspectives as a source of self-evaluation and self-control, the greater their capacity to differentiate roles and extend the scope of social organization (Mead 1934, 322).

In this principle, Mead argued that if the basic profile of norms, values, and beliefs of different groupings in which individuals participate is not contradictory, then differentiation does not lead to conflict and degeneration of social organization. On the contrary, multiple and consistent generalized others allow for functional differentiation of roles and groups, which, in turn, expands the scope (size, territory, and other such variables) of society. Of course, if generalized others are contradictory, then conflictual relations are likely, thereby limiting the extent of social organization.

But much like Durkheim (1893), Mead recognized that symbolic components of culture exist at different levels of generality. Some are highly abstract and cut across diverse groupings, while others are tied to specific groups and organizations. Mead (1934, 264–65) distinguished between "abstract" generalized others and concrete "organized" others to denote this facet of symbolic organization. And like Durkheim, Mead saw that the scope of social organization is limited by how well "abstract others" (values and beliefs, for example) are integrated with more concrete "organized others" (particular norms and doctrines of specific groups, social classes, organizations, and regions, for example). Large-scale social organization, Mead felt, is dependent upon common and highly abstract values and beliefs that are integrated with the specific perspectives of differentiated collectivities. The concept of integration, Mead appeared to argue, involves more than consistency and lack of contradiction; it denotes the fact that the specific generalized others of particular organized collectivities represent concrete applications of the abstract generalized other. The abstract generalized other sets the parameters for less abstract perspectives, thereby assuring not just consistency between the two but also integration where the tenets of each are interrelated.

It is in this recognition that Mead's argument came close to Durkheim's arguments in *The Division of Labor* (1893). We can visualize this similarity in the following proposition:

V. The more actors can simultaneously role-take with a common and abstract perspective, and at the same time role-take with a variety of specific perspectives of particular collectivities that are integrated with the abstract perspective, and the more these integrated perspectives can be a source of self-evaluation and self-control, the greater their capacity to extend the scope of social organization (Mead 1934, 321–22).

As with propositions I–IV, the converse of this fifth proposition can point to some of the conditions producing conflict and change. To the degree that abstract and specific perspectives are not integrated, actors will potentially have different interpretations of situations, and to the degree that they come into contact, the probability for conflict will be increased.

These five principles summarize Mead's vision of the basic properties of social organization. For society to exist at all, actors must be able to

role-take with multiple others and with generalized others. For highly cohesive organization to exist and persist, actors must be able to role-take
with a common generalized other. For somewhat less cohesive but more
differentiated and extended patterns of social organization to be viable,
actors must be able to role-take with multiple, but nevertheless noncontradictory, generalized others. And for large-scale and highly extensive
patterns of organization, actors must role-take with well-integrated abstract and specific generalized others.

CONCLUSION

Much commentary on Mead's work involves debates over such issues as
(a) "what he really meant" by concepts like the " 'I' and 'me,' " "role-
taking," "self," and "generalized other"; (b) the indeterminate vs. deter-
ministic strains in Mead's thought (Blumer 1962); (c) the stability of fluidity of self (Blumer 1969; Kuhn and McPartland 1954; Goffman 1959);
(d) the extent to which emotion is a part of the Meadian motivational
scheme (Swanson 1961; Shibutani 1968); (e) the methodological implications of Mead's philosophy (Denzin 1970; Blumer 1969); and (f) the ability
of Mead's scheme to deal with macrostructural accounts of the social
world (Blumer 1962; R. Turner 1978; 344).

The lists of laws presented in this paper can throw considerable
light on these and related issues and thus, a few observations on the implications of examining Mead as a sociological theorist rather than as a philosopher should be made. First, by viewing Mead's scheme as a series of
abstract and universal laws that state the fundamental relationship between individuals and ongoing patterns of social organization, the theoretical power of Mead's ideas is made explicit. These principles can serve
as axioms in deductive schemes designed to explain events in a particular
empirical context. While many (e.g., Blumer 1969) argue that the use of
Mead's ideas in this way violates Mead's intent and that the indetermi-
nistic nature of action and interaction makes deductive schemes irrelevant to sociological inquiry, such a position does not allow for a full appreciation of Mead's genius. Moreover, it rests on a profound
misunderstanding of deductive theory. The abstract laws of deductive
theory can denote only the fundamental properties, and their connections, of some universe; it is left to those working in specific contexts to
use these laws as explanatory principles for understanding the ebb and
flow of particular events in concrete empirical settings. Deductive theory
does not seek to predict these events; it does not attempt to control for
every variable; it is not primarily concerned with causality. Rather, it
seeks to explain events by demonstrating that they occur in conformity
with certain fundamental processes which are stated as an abstract law or

axiom. In many ways, when Mead's ideas are employed as a metaphor, as metatheory, or as "sensitizing concepts" (Blumer 1969), the logic of deductive theory is implicitly employed, since concrete empirical events are seen to occur in terms of Meadian assumptions. By formalizing Mead's ideas, this implicit deductive tactic is only made more explicit.

Secondly, the issues of indeterminacy, fixity of self, and confusion over concepts are, to some extent, obviated by viewing Mead's scheme as a series of theoretical principles. Mead's laws state that certain properties of the social universe are connected and related to each other. For example, role-taking, self-evaluation, and self-criticism with regard to generalized others are connected to the scope of social organization. This statement says nothing about causality, determinacy, or fixity or fluidity of self, nor does it precisely define self and role-taking. Different scholars with different views on the issues of determinacy and fixity of self and with somewhat different definitions of critical concepts can, it is argued, still use this principle to understand the social world. While it would be preferable to have consensus over definitions and over metatheoretical assumptions, such will probably never be the case for Mead's work, since as verbatim lecture notes it suffers from terminological slippage. More fundamentally, Mead was more interested in denoting basic properties and their relations than in providing precise definitions and an assumptive straitjacket. By formalizing Mead's ideas as a series of propositions, these *general* properties are denoted and their *basic* relations are clearly articulated, without excluding any particular group of scholars who bring to Mead, or pull out of Mead, somewhat different assumptions and definitions. Meadian scholars of many different persuasions should find Mead's principles, as delineated earlier, useful in understanding the social world.

Third, when Mead's ideas are presented more formally, their relevance for understanding macropatterns of social organization is highlighted. Too often, Mead's ideas have been vaguely asserted as relevant to macroprocesses. For example, Blumer proclaims that "society is symbolic interaction" without really specifying the ways in which this is so (Blumer 1962). In contrast, the formal propositions demonstrate the convergence of a macrofunctionalist like Durkheim and the social behaviorism of Mead. Indeed, Mead's scheme specifies the behavioral mechanisms by which extended patterns of social organization are "held together" by cultural ideas. What the propositions underscore is the connection between individual behavioral capacities for role-taking, mind, and self on the one hand, and social structural elaboration on the other. For ultimately, Mead argued, social structures are possible only by virtue of the human capacity to role-take with varieties of concrete and abstract generalized others and to use the perspective of such others as a basis for social control (Shibutani 1955). While separate theoretical principles are required for understanding the form of macrosocial patterns, Mead's principles present a clear picture of the underlying social psychological dynamics of such patterns. By formalizing Mead's principles, then, it is possible to see both the strengths and limitations of Mead's analysis for understanding macrosocial structure.

Finally, returning to my opening arguments about problems in sociological theory, it should be emphasized that Mead's laws come as close as is possible in sociology to being the sociological equivalent to theoretical principles in other scientific disciplines. Sociologists spend a considerable amount of time being defensive because they have not "found" an Einstein, Darwin, or similar figure who unlocked the mystery of a particular universe. For scholars such as Merton (1968, 47), sociology will have to wait for its Einstein because "it has not yet found its Kepler—to say nothing of its Newton, Laplace, Gibbs, Maxwell or Planck." Such pessimism is unfounded, for Mead's laws are the equivalent of those developed by Einstein. Mead unlocked some of the fundamental mysteries of the social universe by articulating the basic laws of action, interaction, socialization, and social organization. While many in sociology do not hold to Comte's advocacy for "social physics" as either possible or desirable, Mead's ideas prove these pessimists wrong. Sociology has theoretical principles equivalent to those in other "hard sciences." However, current sociology is often too mired in statistical and ethnographic descriptions of specific empirical events and too theoretically inhibited by undue metatheoretical agonizing to recognize laws of human action and organization when they are clearly evident to all who will look. Hopefully, the principles presented in this paper will encourage sociologists to take Mead, and other early masters, more seriously as theorists. For sociology has had several "Einsteins," but it has yet to appreciate them.

15 What Went Wrong? Detours from the Early Masters

Sociologists often feel somewhat embarrassed in the presence of "hard scientists." Even within the social sciences, sociologists sometimes appear defensive when comparing their conceptual accomplishments to those of economists and psychologists. Much of this discomfort stems from the lack of mature theory in sociology. Indeed, what pride can we take in our theoretical accomplishments when compared to those in physics and biology? Where are sociology's equivalents to Einsteinian or Darwinian theory? Or, if we come closer to home, where is our theoretical answer to even the admittedly parochial principles of behaviorism or those of classical economics?

Sociologists often take solace in noting the advancement in quantitative data analysis techniques, even though these represent simple elaborations of the correlational analysis developed by a British agronomist, Karl Pearson. While many of these techniques represent number-crunching razzle-dazzle, we can still take pride in the methodological and statistical advancements in sociology. These advancements, however, only throw into bolder relief the lack of well-developed, formal, and interesting theory in sociology. And so, at a time when our theory-testing capacity has taken great leaps, the last fifty years have seen comparatively few major theoretical developments in American sociology.

Such a conclusion is perhaps unduly harsh. One might point to a number of theoretical developments in diverse areas. For example, label-

This chapter was originally titled "Sociology as a Theory Building Enterprise: Detours from the Early Masters" and appeared in the *Pacific Sociological Review* 22 (1979): 427–56.

ing theory, Parsonian action theory, ethnomethodology, symbolic inter-actionism, role theory, critical theory, and exchange theory all contain uniquely American elements. Yet if we examine these theoretical "schools of thought," many are little more than metatheoretical supposi-tion, others are warmed-over versions of the work of the great European thinkers of the last century, and still others represent restatements of Meadian social psychology. There are, of course, exceptions, but theory in contemporary America (and things are even worse in Europe) does not reveal the formality or power of theory in other sciences, nor does it ex-cite, to any great degree, the minds or imaginations of those who are most capable of testing theories.

Why should this be so? Why should Auguste Comte's dream of a "science of society"—indeed, a "social physics"—have come to a state where those of different philosophical and metatheoretical persuasions argue past one another (or, even worse, talk only with those of their kind), where one must go begging to find an abstract theoretical principle, or where many empiricists simply view theory as irrelevant to their desire to crunch numbers in new ways? In a word, what went wrong?

To answer this question, I need to address several issues. First, I must review the various explanations that sociologists offer as a defense, or as an excuse, for their theoretical failings. Second, I must go back to the hundred years before the last sixty—that is, to the period between 1830 and 1930—to discover if sociology "got off on the wrong theoretical foot" or simply lost the inspiration and genius of its founders. For if sociology is to make any pretense of being or becoming a science, it will have to come to grips with the reason for our implicit embarrassment when in the pres-ence of hard scientists. This embarrassment stems from our lack of an accepted body of theoretical principles that allows us to explain why events in the social world occur.

EXPLANATIONS OF SOCIOLOGY'S THEORETICAL FAILINGS

The most frequently cited reason for sociology's theoretical failings is its position as a nascent science. Such a view is really an excuse, because formal sociology is 150 years old, and because humans have been think-ing about themselves and their condition for centuries. As long as sociol-ogy remains comfortable with this view, it will grow old while remaining scientifically immature.

A second reason for sociology's failings comes from those who be-lieve that sociology has an insufficient data base for inducing theory or testing its implications. This view is based on a false conception of theo-retical activity, for, in reality, much theory is generated without intimate knowledge of "the facts." Moreover, those who are buried in mounds of data are rarely able to abstract above those facts. Indeed, conceptual and theoretical skills are much different—and require different mental processes—from data-gathering skills. Thus, if sociology waits for the accumulation of more "facts," it will continue to inspire new data-analysis techniques, but it will thwart its development as a science organized around basic theoretical principles.

A third explanation for sociology's failings comes from those, such as Merton (1968), who argue for middle-range theories as the necessary prerequisite for more general theoretical principles. This advocacy has, since the demise of Parsons's grand intellectual scheme, dominated sociology. Unfortunately, the middle-range strategy has not been implemented in the way Merton intended. Rather than theories of limited range, in terms of their level of abstraction and breadth of coverage, we have generated a series of highly specific theories in a number of diverse substantive areas that are, in many ways, little more than collections of generalizations of empirical findings. Instead of well-developed theories of generic and basic processes, such as conflict, cooperation, socialization, accommodation, assimilation, dissociation, and the like, and instead of theories of basic types of social structures—such as hierarchies, ecological systems, communities, organizations, or groups—we have "theories" of family, criminal gangs, finance units in organizations, economic development, ethnic minorities, and the like. These are not middle-range theories; they are places where one might test a theory. To paraphrase Homans (1961), they are "where one studies, not what one studies." Thus, the middle-range strategy has created a series of interesting empirical generalizations, typically presented as causal models or some other correlational devices, *as if* they were theory. As a result, sociology often confuses the procedures for testing theories with the process of constructing theory.

Still another explanation of why sociology produces so little theory comes from diverse camps, all of which view the "natural science" conception of theory as inapplicable to analyzing the social world. There are those who believe that human behavior and organization contain the capacity for spontaneity and indeterminism, with the result that there are no timeless or universal patterns of organization describable in terms of abstract laws (Blumer 1969); still others argue that each historical epoch reveals its own laws of organization, thus rendering the search for panhistorical or universal laws fruitless (Appelbaum 1978). Another group of scholars believes that the methodological problems of humans studying humans are so great as to make deductive theory and its definitive refutation a virtual impossibility—or at least that these facts require as a first priority the discovery of the laws of human thought, cognition, and consciousness, because all knowledge about patterns of social organization is mediated through such mental processes (Cicourel 1973, 1964).

These explanations arrest our theoretical imagination. They imply that abstract theory cannot be developed for social phenomena; or if it can, it must wait for a more adequate data base, a body of middle-range theories, a prior theory or philosophy of cognition, and worst of all, an unspecified number of years until intellectual maturity sets in. These explanations are feeble and incorrect. Theory in other sciences has often come early in the history of a discipline; it has frequently come without extensive catalogues of facts; and it has had to overcome methodological obstacles equal to those in the social sciences. (We could view the fact that we are humans studying humans as an *advantage* rather than a handicap, because we can intuitively achieve familiarity with our data.)

If we go back to the beginnings of sociology, we can find little trace of current methodological inhibitions. Indeed, the first sociologists were filled with optimism in the post–Newtonian era. They believed that universal laws of human organization could be discovered, and they acted on this belief. The result was that, by 1930, sociology had a legacy of a hundred years of bold and venturesome theoretical activity by men who were not intimidated by the complexity of phenomena, who recognized that no theory can control for every variable, and who were willing to guess and be proven wrong. While we still revere Marx, Durkheim, Mead, and a few others, we do not look at them in the way that they looked at themselves or at each other. We view them through the eyeglasses of our theoretical inhibitions, with the result that we do not ask: What are the universal theoretical principles developed by our first masters? What was our theoretical legacy in 1930? Rather, we trace the connections between schools and the "intellectual milieu of their times" (Coser 1977; Nisbet 1966). We debate Durkheim's "real meaning" of the concept of anomie (Lukes 1973). We argue over how much Weber reacted to "the ghost of Marx" (Bendix 1968). We debate the fine points and nuances of Marx, much as pedantic religious scholars pore over sacred texts (Hook 1974; Ollman 1976). In other words, we concern ourselves primarily with distracting details while we ignore the essence of any science—the development of abstract theoretical principles.

The profound tragedy of all this concern with scholarly minutiae is that many giants of our past had developed some basic theoretical principles that we ignore. Why have we ignored the important part— that is, the theoretical principles—of our first one hundred years? Part of the answer resides in our fear of abstraction; another part comes from our current obsession with causality; yet another part stems from our desire to find statistical significance on a subject matter that does not matter; and, most tragically, a large part of the answer results from our political bias and the blind rejection of scholars whose politics do not correspond with our own.

If we are willing to reject current explanations of sociology's theoretical shortcomings, then we must ask: Where did sociology go wrong? What happened to distract the sociological imagination? The answer to these questions lies in our first one hundred years as a formal discipline; and, thus, we need to examine the theoretical strategies, as well as the abstract principles, of our early masters in order to see what they presented to us by 1930. Only in this way can we understand the theoretical events of the last sixty years in American sociology.

THEORETICAL STRATEGIES OF THE EARLY MASTERS, 1830–1930

In table 15.1 the theoretical strategies of sociology's most prominent early masters are presented. In particular, I have emphasized the views of Comte, Spencer, Marx, Durkheim, Weber, Pareto, and Mead with respect to the following strategic considerations:

1. *Abstract Principles.* Should, or can, sociological theory resemble theory in physics? Can we develop abstract statements of fundamental relations among social phenomena, without regard to causality and without concern for historical epochs?

2. *Causality.* To what extent should sociological theory be concerned with discerning the causes of phenomena? Must sociology's abstract statements uncover causal connections?

3. *Typology and Classification.* To what degree should sociological theory employ abstract or concrete typologies of social phenomena? Should these be constructed prior to theoretical statements?

4. *Structural Affinities.* Should sociological theory be devoted to understanding the covariance of specific types of social structures (such as how changes in the economy are associated with changes in kinship and religion)? Or should theory seek to understand the more abstract properties of social interaction and organization common to all types of social structures (such as the principles of hierarchy, ecological distribution, mobility, and the like)?

5. *Metatheoretical Supposition.* To what extent must sociology begin with a series of assumptive statements on the "nature of social reality" before theory can be developed? Can theory be developed only after creating an extensive metaphysical system?

6. *Induction versus Deduction.* Should detailed observations of social events precede and inspire the development of theory? Or should theory be articulated first, and then tested against the empirical facts?

TABLE 15.1
Diverse Theoretical Strategies of the Early Masters

	Abstract Laws	Degree of Emphasis on					Induction vs. Deduction
		Causality	Typologies	Structural Affinities	Metatheory		

	Abstract Laws	Causality	Typologies	Structural Affinities	Metatheory	Induction vs. Deduction
Auguste Comte (1798–1857)	Sociology can emulate physics and seek invariant laws of the social universe.	Concern with causality will detract from the search for laws.	Useful in describing stages of societal development.	Useful in showing how structures change together during societal development.	Organismic analogy: Social phenomena reveal systemic properties.	Theory must be based on observation, and vice versa.
Herbert Spencer (1820–1903)	Sociology must seek the laws of the social universe, as these can be deduced from the law of cosmic evolution.	Should be concerned with causality, but must be secondary to search for laws.	Useful in describing stages of societal development and in capturing the cyclical dynamics of social systems.	Useful in creating a data base for inducing and testing theories. Engaged in a lifelong effort to describe social structures of diverse types of societies.	(1) Organismic analogy: Social phenomena reveal systemic properties. (2) Implicit functionalism: Appropriate to analyze needs of the social system met by a particular structure.	Theory must be based on observation, and vice versa.
Karl Marx (1818–1883)	Laws of distinctive historical epochs can be discovered and used to analyze events in that epoch. Universal laws for all times and places cannot be discovered.	Must be concerned, since laws will be causal statements of economic determinism.	Not as important as laws that describe the dynamics of a historical epoch.	Will reflect causal connections. Can be used to discover laws of each epoch.	(1) Dialectics: Structures contain the very properties that lead to their transformation. (2) Conflict-change: Change is the result of conflict among super- and subordinate classes. (3) Sub-superstructure: Economic variables determine cultural and social patterns.	Laws of a historical epoch can be induced from observations of the relations of production, and/or through praxis.

Max Weber (1864–1920)	Not concerned.	Should trace the causal relations among social phenomena.	Typologies, or ideal types, are the essence of sociological description.	The goal of sociology is to show how empirical structures covary and, if possible, are causally connected.	(1) Action is meaningful and must be understood at this level. (2) Action creates emergent patterns that are amenable to sociological analysis.	Conceptual work is to be induced from a careful examination and comparison of empirical cases.
Émile Durkheim (1858–1919)	Regularities in human organization can be articulated.	First and final causes must always be assessed—that is, the antecedent conditions and functions of a phenomenon must be determined.	Useful in capturing the variable states of phenomena.	Necessary to discovering social patterns. Few such affinities actually articulated.	(1) Organismic analogy: Phenomena reveal systemic properties. (2) Explicit functionalism: Necessary to determine the integrative needs served by social phenomena.	Observations on empirical and historical events to form the basis for causal and functional statements.
Vilfredo Pareto (1848–1923)	Sociology can emulate physics and discover the invariant laws of social organization and change.	Analysis of one-way causality will inhibit sociological inquiry. Must focus on mutual connections of phenomena.	Useful in describing variable states of phenomena.	Social phenomena vary together, and hence, empirical descriptions of the patterns of covariance are critical to sociological analysis.	(1) Social phenomena reveal equilibrium tendencies. (2) Change reveals cyclical patterns.	Theory must be based upon observations, and vice versa.
Georg Simmel (1858–1918)	Laws of association can be discovered. Little concern with explicit articulation.	Not concerned.	Not concerned.	Necessary to unraveling the basic forms of social interaction. Few actually developed.	Social phenomena reveal underlying forms that can be described.	Implicit inductive emphasis.
George Herbert Mead (1863–1931)	The fundamental nature of the relationships between individuals and patterns of social organization can be articulated. Little concern with formal laws, however.	Not concerned.	Not concerned.	Not concerned, except to show that mind, self, and society are interrelated.	(1) Mind and self are behaviors. (2) Social organization cannot exist without mind and self, and vice versa.	Advocated neither, but scheme is implicitly deductive.

Of course, these issues are not mutually exclusive, but concentration on some as opposed to others will have profound consequences for the kind of theory which will be articulated. If one scholar believes, for example, that a well-developed set of metatheoretical assumptions must precede the articulation of abstract principles, whereas another insists that typologies must precede causal statements, the form, style, and substance of their respective theories will differ—as would the theories of scholars emphasizing different combinations of these five strategic issues.

Table 15.1 allows us to compare the diversity of strategies of our first hundred years with those of the last fifty. In general, the first self-conscious sociologists, Auguste Comte and Herbert Spencer, advocated a sociology based on the Newtonian vision of science. This strategy emphasized the search for the laws of social organization and change, without overconcern with causality. Comte, in particular, saw a concern with causality as hindering sociological theory (1830–1842b, 5-6):

> The first characteristic of Positive Philosophy is that it regards all phenomena as subject to invariable natural *Laws*. Our business is,—seeing how vain is any research into what are called *Causes*, whether first or final,—to pursue an accurate discovery of these Laws, with a view to reducing them to the smallest possible number. By speculating upon causes, we could solve no difficulty about origin and purpose. Our real business is to accurately analyze the circumstances of phenomena, and to connect them by the natural relations of succession and resemblance. The best illustration of this is in the case of the doctrine of Gravitation. [emphasis in original]

Spencer did not hold such an extreme view, although his concern was always with exposing the "relations of affinity" of social phenomena. Despite modern commentators' misplaced emphasis, neither Comte nor Spencer placed a heavy emphasis on typologies. True, Comte (1830–1842a) articulated a "law of three stages" and Spencer (1874–1896) presented two basic typologies, one distinguishing "militant" and "industrial" social forms and the other the stages of social evolution from "simple" to "compound," through "doubly compound," and to "trebly compound." Such typologies were used to describe certain structural affinities during the course of social structural evolution; and, in fact, Comte and Spencer were far more interested in describing the shifting relations among basic economic, political, family, community, legal, and religious structures during societal evolution than in elaborating taxonomies. Such structural regularities, both felt, were evidence of the operation of universal and invariant laws. With respect to metatheoretical considerations, Comte and Spencer employed the "organismic analogy" to emphasize the systemic character of social phenomena; and contrary to many commentators, neither Comte nor Spencer viewed "society as an organism." In regard to the issue of induction versus deduction, neither Comte nor Spencer would phrase the issue in this way. As Comte (1830, 42) emphasized, "If it is true that every theory must be based upon observed facts, it is equally true that facts cannot be observed without guidance of some theory." Or, as Spencer emphasized in the introduction to his often-ignored *Descriptive Sociology* (1873–1934) and throughout *Principles of Sociology* (1874–1896), theory and observation must be in constant interaction, with each checking the other. But both emphasized that idle observation, or "raw empiricism," would inhibit the development of sociology. As Comte (1830, 242) stressed:

> The next great hindrance to the use of observation is the empiricism which is introduced into it by those who, in the name of impartiality, would interdict the use of any theory whatever. No longer dogma could be more thoroughly irreconcilable with the spirit of the positive philosophy. . . . No real observation of any kind of phenomena is possible, except in as far as it is first directed, and finally interpreted, by some theory.

Thus, in sociology's theoretical beginnings, there was an initial emphasis on the search for abstract laws under the minimal metatheoretical assumption that social phenomena reveal systemic properties. Concern for causality and typologies was viewed as secondary to the goal of articulating abstract principles. Empirical descriptions revolving around structural affinities during societal development were to be the data base for testing, and at times inducing, abstract theoretical statements. From this initial strategy, however, sociological theory was to diverge in the works of Marx, Durkheim, and Weber. Pareto and Simmel—in particular, Pareto—retained some of this early vision but it was Durkheim, Weber, and Marx who guided sociological theory in the early twentieth century away from its early strategy. This divergence from the early strategy of Comte and Spencer is, as we will see, one of the major reasons for the lack of well-developed sociological theory in sociology.

If one reads down the columns and across the rows of table 15.1, this shift in theoretical strategies becomes clear. Those scholars who are most revered in contemporary theoretical circles became increasingly concerned with causality, typology, and descriptions of structural affinities. Thus, Durkheim's concern with cause in *The Division of Labor* (1893) and *Suicide* (1897), as well as in his explicit advocacy of functional analysis (1895), redefined Comte's positivism in such a way that the concern for laws recedes, and the very thing that Comte feared most—a concern for "first" and "final" causes—resurfaces. Weber's approach similarly advocated the importance of causality, best exemplified in *The Protestant Ethic and the Spirit of Capitalism* (1905). Moreover, under Weber, theory became increasingly taxonomic and classificatory, emphasizing ideal types and the discovery of the "place" of a phenomenon in relation to an "ideal" or "pure" form (Shils and Finch 1949). Marx's advocacy was also to arrest sociology's concern for universal abstract principles, with its advocacy of laws of historical epochs. As a philosopher, Mead (1934) was concerned not with theory per se, but with discovering fundamental truths—a concern that could have become the basis for abstract theoretical principles, except for the fact that many interpreters of Mead, such as Blumer (1969), took his thought in a more theoretical direction. Those scholars who advocated strategies that were compatible with Comte's vision—men such as Pareto and Simmel—were ignored in the early decades of this century. When their work was finally introduced, it came in overly functional trappings.

The result of this shift in theoretical strategy has been for theory to be dominated by an overconcern with tracing causal connection among specific empirical phenomena. It was during this period that sociology became the repository for "theories of _____" (fill in a substantive area). Such "theories" have been useful for applied and practical purposes, but all too often they are insufficiently abstract and too narrow to be the organizing principles of a science. Only in the last decade or so has there been a renewed interest in developing more abstract principles that cut across and subsume these more specific generalizations of various substantive fields. This rediscovery of Comte's and Spencer's early advocacy needs to be assessed in relation to the theoretical legacy of sociology's first one hundred years.

THEORETICAL PRINCIPLES OF THE EARLY MASTERS

The Misguided Reverence for the Masters

One of the problems with developing abstract theory in sociology is the tendency to evaluate theory in terms of sociology's current obsession with multivariate analysis. The goal of multivariate analysis is to "explain" degrees of variance in some empirical phenomena. Concern is with statistical controls and the introduction of additional variables in an effort to increase the amount of variance explained. These concerns underscore the descriptive nature of multivariate techniques, but unfortunately the "models" constructed from these techniques are often viewed as theory. In point of fact, these models are descriptions of empirical events in particular contexts at particular points in time. They rarely model generic properties of all social systems in all times and places, with the result that they are not sufficiently abstract to constitute laws of human organization.

Yet we tend to view with suspicion any theoretical activity that seeks to reduce the number of variables, that does not attempt to control for every specific empirical condition, that ignores causal connections, and that seeks only to state the fundamental affinities of generic social processes. Indeed, sociologists are likely to view such theoretical activity as "simplistic," as too "detached from the empirical world," as "untestable," and so on. But theory in the respected sciences is simplifying, detached, and often untestable when first formulated (and, at times, not directly testable at all). Thus, even if sociology had its Einstein, it is likely that the sociological equivalent of $E = mc^2$ would be criticized as too abstract and simplistic, lacking "operational definitions," as ignoring the significance of a plethora of empirical variables; and it would be attacked for ignoring causality. We would, one suspects, not recognize sociology's Einstein. Such has been the "progress" of sociology over the last fifty years.

In many ways, it can be argued that we have had several "sociological Einsteins" and have simply not fully appreciated their importance. Spencer, Marx, Durkheim, Pareto, Simmel, and Mead did indeed unlock many of the basic mysteries of the social universe, and they did articulate as abstract laws some of the fundamental processes in this universe. Over the last sixty years modern sociology has given considerable credit to these giants of our past; but unfortunately we sometimes do not take these figures seriously as theorists—that is, as scholars who saw at least some of the basic properties and processes of the world and who articulated some of sociology's basic laws. Rather, we tend to revere them for the wrong reasons—Durkheim for his causal analysis, Weber for his descriptive-analytical powers, Marx for his dialectics, Pareto for his concept of "nonlogical action," Mead for his philosophy of self, and so on. Or we debate the "real meaning" of concepts such as "anomie," "egoism," "alienation," "class," the "I and me," and so on for virtually any term used by the first masters.

Thus, we have often been both uncritical and atheoretical in our analysis of the masters. One way to begin a reassessment of these masters is to examine some of the basic laws of human organization that they uncovered. These laws denote certain properties of the social world as basic and generic—that is, as inherent in human organization in all times and places. From the viewpoint of our current multivariate mania, these laws may seem too simplistic, excessively abstract, and unconcerned with causality. Yet these are the characteristics of theory as opposed to current empirical descriptions that masquerade as theory.

Some Laws of Human Organization and Interaction

One way to assess accomplishments in American sociological theory for the last fifty years is to compare current efforts with the cumulative legacy of the most important scholars listed in table 15.1. In summarizing this legacy, as it stood in 1930, I intend to answer a hypothetical question: What if Comte's early advocacy had been followed? What would sociology's theoretical legacy look like if it were to be translated into abstract theoretical principles? While Marx, Durkheim, or Mead might complain about such an exercise as it related to their work, it is likely that it would have been performed by their contemporaries *if* Comte's version of positivism had held sway beyond 1850, or *if* Spencer's advocacy had stimulated students of theory. Thus, what is being proposed is merely a belated effort at what should have been done many decades ago.

What basic properties of the social world did the first masters perceive as most generic and fundamental to human organization? If we approach the masters with this question, four basic processes appear as fundamental to the social world: (1) differentiation, (2) integration, (3) disintegration, and (4) interaction. Attention to the first three represented the European contribution, while focus on the last was distinctly American. The most abstract principles formulated with respect to these processes are summarized below.

The Process of Differentiation

Whether the scholar be Spencer, Marx, or Durkheim, social thinkers all recognize that differentiation of social units in social systems represents a basic property of the social world. And while different scholars studied diverse aspects of this process, their combined legacy, when viewed at the most abstract level, can be summarized as three basic laws of social differentiation:

I. The greater the degree of productivity in a social system, the greater the level of differentiation in that system (Marx 1848, 1867; Spencer 1874–1896).
   A. The level of productivity is a positive function of the availability of relevant resources (Spencer 1874–1896) and the level of relevant technology (Marx 1867).
II. The greater the level of competition for resources in a social system, the greater the level of differentiation in that system (Spencer 1874–1896, 1862; Durkheim 1893).
   A. The level of competition is a positive function of absolute population size (Spencer 1864–1867, 1874–1896), rate of population

growth (Spencer 1864–1867; Durkheim 1893), and degree of ecological concentration of a population (Durkheim 1893).

III. The greater the degree of differentiation in a social system, the greater the degree of differentiation along functional, ecological, and rank dimensions.

A. Functional differentiation will initially occur along the regulatory and productive axes, and only later along the distributive axis (Spencer 1874–1896, 1864–1867).

B. Ecological differentiation will initially occur along the productive axis, and only later along the regulatory and distributive axes (Spencer 1874–1896).

C. Rank differentiation will initially occur along the productive and regulatory axes, and only later along the distributive axis (Marx 1867, 1848).

The first principle states that productivity and differentiation are fundamentally related; the second argues that competition and differentiation are related; and the third principle denotes the nature, form, and direction of differentiation. This last principle requires some elaboration. Borrowing from Spencer (1874–1896), three functional axes of differentiation are distinguished, with the "regulatory" pertaining to those units involved in controlling internal processes and relations with the external environment, with the "productive" denoting internal processes that generate the substances on which system units persist, and with "distributive" concerning the flow of materials and information within the system. From Spencer, but also Marx (1867) and, to a lesser extent, Durkheim ([1893] 1922), differentiation also is seen to have an ecological dimension, with the productive axis dictating the ecological distribution of system units. From Marx (1867, 1848) and to some extent from Spencer (1874–1896), Durkheim (1893), and Pareto (1916), differentiation is considered to have a hierarchical dimension, with the productive axis being initially most related to its form.

Several observations should be made on these and subsequent principles. First, they are highly abstract, as any law of social organization must be. While Marx, Spencer, Durkheim, and others tended to focus on societal social systems, their principles can apply to other types of systems—groups, organizations, communities, and the like. Second, the principles are considered to be true for all times and places, although specific empirical conditions and events will influence the weights of the variables (but will not obviate the fundamental relationship). Third, these principles are concerned not with causality (this is an empirical question), but with establishing that certain properties of the social universe—patterns of differentiation, productivity, competition, for example—are fundamentally related. Such a concern is what Comte had in mind when he emphasized that "our real business is to analyze accurately the circumstances of phenomena, and to connect them by natural relations of succession and resemblance (Comte 1830–1842, 6)." While such principles may seem obvious, simplistic, and nonempirical through the prism of multivariate mania, they are in form (and, most certainly, content) what the laws of sociology will look like. Their empirical implications can be studied with multivariate techniques as deductions are made to concrete empirical systems, whether nation-states (Nolan 1979), administrative bureaucracies (Blau 1974), or communities (Hawley 1950). But these empirical implications, and the multivariate techniques that tease them out, are *not* theory. The theory must be highly abstract and state the nature of relations among only a few generic variables. This fact is too often ignored in sociological theorizing—again, one of the unfortunate events of the last sixty years in American sociology.

The Process of Integration

All scholars of the nineteenth century were vitally concerned with the processes of integration in society. Despite the surface differences in their analyses, three laws of social integration in differentiating social systems emerge from their collective work:

1. The greater the degree of differentiation in a social system, the greater the degree of centralization of regulatory processes in that system (Spencer 1874–1896; Pareto [1916] 1935; Durkheim [1893] 1933).
   a. The degree of centralization is a positive function of the degree of external (or perceived) threat to a system (Spencer 1874–1896; Simmel [1908] 1956) and the degree of dissimilarity of system units (Spencer 1874–1896; Marx and Engels [1848] 1955).
2. The greater the level of structural differentiation in a social system, the greater the degree of mutual dependence and exchange of resources among differentiated system units (Spencer 1874–1896; Durkheim [1893] 1933).
3. The greater the level of structural differentiation in a social system, the greater is the degree of generalization of evaluational cultural symbols, the greater is the degree of specificity of norms within and between system units, and the greater is the degree of consolidation of similar specialties into collective units (Durkheim [1893] 1933).

These principles stress that in differentating social systems, integration, centralization, mutual interdependence, and exchange, generalization of values, normative specification, and subgroup formation are fundamentally related in the social universe. As with the principles of differentiation, the presumption is that they apply to all differentiating systems and that they transcend particular historical epochs. Moreover, as deductions to specific contexts are made, empirical variables will be added as the unique features of situations are encountered (although the postulated relations are unaltered).

The Process of Disintegration

The process of integration involves the structuring of relations among differentiating units, whereas the process of disintegration denotes the destructuring of such relations. All major theorists of the last century viewed human organization as revealing an inherent dialectic between processes of integration and processes of disintegration. Indeed, a separate theory of change and stasis was not required, because social change and stability reflected the balance between these two basic processes— an analytical insight often ignored in American sociology over the last sixty years. At the most abstract level, three laws of disintegration are evident in the work of the first masters.

1. The greater the degree of generalization of evaluative cultural symbols without compensating normative specification and/or organization of specialties into collective units, the higher the rates of deviance in that system and the more prevalent are disintegrative processes (Durkheim [1897] 1951 [1893] 1933).
2. The greater the degree of inequality in the distribution of scarce resources in a differentiating social system, the greater the degree of resistance by those segments low in resource shares in that system and the more prevalent the disintegrative processes (Marx and Engels [1848] 1955; Spencer 1874–1896; Pareto [1916] 1935).
   a. The degree of resistance is a positive function of the degree of centralization of power (Pareto [1916]; Spencer 1874–1896; Marx and Engels [1848] 1955).
   b. The degree of resistance is a positive function of the degree of rank differentiation (Marx and Engels [1848] 1955).
3. The lower the degree of centralization of regulatory processes in a differentiating social system, the lower the capacity for control and coordination in that system and the more prevalent are disintegrative processes (Durkheim [1893] 1933; Spencer 1874–1896; Pareto [1916] 1935).

The first proposition states Durkheim's argument on "anomie" and "egoism" (1893, 1897). The second proposition incorporates two distinctive lines of thought—(1) Pareto's, Marx's, and Spencer's recognition that centralized power (one type of resource inequality) inevitably generates resistance and counterpower; and (2) Marx's arguments on class formation and conflict. The third proposition emphasizes that decentralization decreases integration at the point where regulation, coordination, and control of units breaks down. Thus, both centralization and decentralization of power reveal inherent disintegrative processes. These propositions allow for understanding of disintegrative (and, hence, change) processes in many types of systems; and while specific empirical conditions will determine the weights of the variables, the postulated relations among them remain unaltered.

The Process of Interaction

With the possible exception of Simmel, European theory had been predominantly macrostructural in its emphasis; and when it did seek to understand the connections between the individual and society, it was noticeably inadequate. Indeed, it is painful to read Spencer's utilitarianism. Marx's psychological ramblings, Pareto's discussion of "sentiments" and "residues," Durkheim's late fumblings on the mind, and Weber's advocacy and abandonment of "action." Only in the work of Mead (1934), who synthesized the work of others, was the fundamental nature of human interaction, and its connection to ongoing patterns of social organization, to be captured. Indeed, in Mead's work a series of implicit principles grasp the fundamental nature of individual interaction and social organization.

  I. The more an actor must consummate impulses in organized social
     contexts, the more developed are the behavioral capacities for role-
     taking, mind, and self (Mead 1934).
 II. The more an actor can role-take with another actor and use the in-
     formation thus gained as a source of self-control and self-evaluation,
     the greater is that actor's capacity to interact and cooperate with
     others (Mead 1934; Simmel 1956).
III. The more an actor can role-take simultaneously with multiple others
     and generalized perspectives and use the information thus gained as
     a source of self-control and self-evaluation, the greater is that actor's
     capacity to interact and cooperate with others in diverse, complex,
     and extended patterns of social organization (Mead 1934).

In these principles, Mead argued that the acquisition of the crucial capacities for role-taking, mind, and self (proposition I) occurs primarily in accordance with the principles of behaviorism, since mind, self, and role-taking bring about increased adjustment and adaptation to organized social contexts (the ultimate reinforcer). Propositions II and III document Mead's and, to a lesser extent, Simmel's recognition that patterns of social organization are possible only by virtue of role-taking with another actor, multiple others, and varieties of "generalized others" and the use of the information from these various types of others for self-evaluation and self-control. While Mead enumerated many more specific principles, these three are the most abstract and document the fundamental relationship in the social universe between the behavioral capacities of individuals for mind, self, and role-taking, on the one hand, and extended patterns of social organization, on the other.

In sum, these twelve principles, and many more specific ones that could be derived, on the processes of differentiation, integration, disintegration, and interaction formed in 1930 a solid foundation for sociological theorizing over the last sixty years. If these and related laws could have served as a *starting* point for theorizing during the last decades, sociological theory would be far more advanced; and sociology as a scientific enterprise would be more respected by the hard sciences. And Comte's dream of a social physics—that is, a sociology organized around the fundamental laws of the universe—would be much closer to realization.

References

Alexander, Jeffrey C. 1982. *Theoretical Logic in Sociology*. vol. 2. Berkeley: University of California Press.

Almond, Gabriel, and Sidney Verba. 1965. *The Civic Culture*. Boston: Little Brown.

Appelbaum, R. P. 1978. "Marx's Theory of the Falling Rate of Profit." *American Sociological Review* 43 (February): 67–80.

Baldwin, John D. 1986. *George Herbert Mead: A Unifying Theory for Sociology*. Newbury Park, Calif.: Sage.

Banfield, Edward. 1958. *The Moral Basis of a Backward Society*. New York: Free Press.

Baran, Paul, and P. M. Sweezey. 1966. *Monopoly Capital*. New York: Monthly Review Press.

Bendix, Reinhard. 1960. *Max Weber: An Intellectual Portrait*. Garden City, N.Y.: Doubleday.

_____. 1968. *Max Weber: An Intellectual Portrait*. Garden City, N.Y.: Doubleday.

Berger, Joseph, David G. Wagner, and Morris Zelditch. 1988. "Growth, Social Processes, and Meta Theory." In *Theory Building*, edited by J. H. Turner. Newbury Park, Calif.: Sage.

Blalock, Hubert M. 1967. *Toward a Theory of Minority Group Relations*. New York: Wiley.

_____. 1969. *Theory Construction: From Verbal to Mathematical Formulations*. Englewood Cliffs, N.J.: Prentice-Hall.

Blau, Peter M. 1964. *Exchange and Power in Social Life*. New York: Wiley.

_____. 1970. "A Formal Theory of Differentiation in Organizations." *American Sociological Review* 35 (April): 201–18.

_____. 1974. *On the Nature of Organizations*. New York: Wiley.

_____. 1977. *Inequality and Heterogeneity, a Primitive Theory of Social Structure*. New York: Free Press.

Blau, Peter M., and Otis Dudley Duncan. 1967. *The American Occupational Structure*. New York: Wiley.

Blumer, H. 1962. "Society as Symbolic Interaction." In *Human Behavior and Social Process*, ed. A. M. Rose. New York: Houghton Mifflin.

————. 1969. *Symbolic Interaction: Perspective and Method*. Englewood Cliffs, N.J.: Prentice-Hall.

Boulding, Kenneth E. 1962. *Conflict and Defense: A General Theory*. New York: Harper & Row.

Carniero, Robert L. 1967. "Editor's Introduction." *The Evolution of Society*. Chicago: University of Chicago Press.

Childers, Grant W., Bruce H. Mayhew, and Louis W. Gray. 1971. "System Size and Structural Differentiation: Testing a Baseline Model of the Division of Labor." *American Journal of Sociology* 76 (March): 813–30.

Cicourel, A.V. 1964. *Method and Measurement in Sociology*. New York: Free Press.

————. 1973. *Cognitive Sociology*. London: Macmillan.

Collins, Randall. 1975. *Conflict Sociology: Toward an Explanatory Science*. New York: Academic Press.

————. 1985. *Three Sociological Traditions*. Oxford, England: Oxford University Press.

————. 1986a. "Interaction Ritual Chains, Power and Property." In *The Micro-Macro Link*, edited by Jeffrey C. Alexander, 177–92. Bernhard Giesen, Richard Münch, and Neil J. Smelser. Berkeley: University of California Press.

————. 1986b. *Weberian Sociological Theory*. Cambridge, England: Cambridge University Press.

————. 1988. *Theoretical Sociology*. San Diego, Calif.: Harcourt Brace Jovanovich.

Comte, Auguste. 1830–1842a. *Cours de philosophie positive, les préliminares généraux et la philosophie mathématique*. 5 volumes. Paris: Bachelier.

————. [1830–1842] 1854. *Course of Positive Philosophy*, 3 vols. Condensed by H. Martineau in 1854. London: Bell & Sons.

————. 1851–1854. *Systéme de politique: Ou, traité de sociologies, instituant la religion de l'humanité*. Paris: L. Mathias.

Cook, Karen S., ed. 1987. *Social Exchange Theory*. Newbury Park, Calif.: Sage.

Coser, Lewis A. 1956. *The Functions of Social Conflict*. New York and London: Free Press.

————. 1967. *Continuities in the Study of Social Conflict*. New York: Free Press.

————. 1977. *Masters of Sociological Thought*. New York: Harcourt Brace Jovanovich.

Cutright, Phillip, and James A. Wiley. 1969–1970. "Modernization and Political Representations: 1927–1966." *Studies in Comparative International Development* 5:23–44.

Dahl, Robert. 1971. *Polyarchy*. New Haven, Conn.: Yale University Press.

Dahrendorf, Ralf. 1957. *Class and Class Conflict in Industrial Society*. Stanford, Calif.: Stanford University Press.

————. 1958. "Toward a Theory of Social Conflict." *Journal of Conflict Resolution* 2 (June): 170–83.

————. 1959. *Class and Class Conflict in Industrial Society*. Stanford, Calif.: Stanford University Press.

Darwin, Charles. [1859] 1890. *On the Origin of Species*. London: Murray.

Davies, James C. 1962. "Toward a Theory of Revolution." *American Journal of Sociology* 27:5–19.

Denzin, Norman K. 1970. *The Research Act: A Theoretical Introduction to Sociological Methods*. Chicago: Aldine.

de Tocqueville, A. [1835] 1969. *Democracy in America*. New York: Doubleday.

Dewey, John. 1922. *Human Nature and Conduct*. New York: Holt.

Dubin, Robert. 1969. *Theory Building*. New York: Free Press.

Durkheim, Émile. 1892. *Montesquieu and Rousseau*. Ann Arbor: University of Michigan Press.

————. [1893] 1933. *The Division of Labor in Society*. New York: Free Press.

————. [1895] 1938. *The Rules of the Sociological Method*. New York: Free Press.

————. [1897] 1951. *Suicide*. New York: Free Press.

————. [1904] 1933. "Preface to the Second Edition." *The Division of Labor in Society*. New York: Free Press.

————. [1912] 1965. *Elementary Forms of Religious Life*. New York: Free Press.

————. [1922] 1961. *Moral Education*. Translated by E. K. Wilson and H. Schnurer. New York: Free Press.

Durkheim, Émile, and Marcel Mauss. [1903] 1963. *Primitive Classification*. London: Cohen & West.

Dye, Thomas R. 1983. *Who's Running America? The Reagan Years*. 3d ed. Englewood Cliffs, N.J.: Prentice-Hall.

Emerson, Richard. 1972. "Exchange Theory, Part 2." *Sociological Theories in Progress*, vol. 2, edited by J. Berger, M. Zelditch, and B. Anderson. Boston: Houghton-Mifflin.

————. 1976. "Social Exchange Theory." In *Annual Review of Sociology*, edited by A. Inkeles and N. J. Smelser, 335–62. Palo Alto, Calif.: Annual Reviews.

Farb, Peter. 1978. *Humankind*. Boston: Houghton Mifflin.

Fink, C. F. 1968. "Some Conceptual Difficulties of Social Conflict." *Journal of Conflict Resolution* 12 (December): 412–60.

Freud, Sigmund. [1913] 1938. *Totem and Taboo*. London: Penguin.

Furfey, Paul Henry. 1953. *The Scope and Method of Sociology: A Metasociological Treatise*. New York: Cooper Square.

Gibbs, Jack. 1972. *Sociological Theory Construction*. Hinsdale, Ill.: Dryden Press.

Giddens, A. 1971. *Capitalism and Modern Theory: An Analysis of the Writings of Marx, Durkheim, and Max Weber*. Cambridge, England: Cambridge University Press.

————. 1972. *Émile Durkheim: Selected Writings*. Cambridge, England: Cambridge University Press.

Goffman, Erving. 1959. *The Presentation of Self in Everyday Life*. New York: Doubleday.

————. 1967. *Interaction Ritual*. New York: Doubleday.

————. 1974. *Frame Analysis*. New York: Harper & Row.

Gurr, Ted. 1970. *Why Men Rebel*. Princeton, N.J.: Princeton University Press.

Hage, Jerald. 1972. *Techniques and Problems of Theory Construction in Sociology*. New York: Wiley.

Hage, Jerald, and Michael Aiken. 1967. "The Relationship of Centralization to Other Structural Properties." *Administrative Science Quarterly* 12:72–92.

Hage, Jerald, Michael Aiken, and Cora Bagley Marrett. 1971. "Organization Structure and Communications." *American Sociological Review* 36 (October): 860–71.

Hannan, Michael T., and John Freeman. 1977. "The Population Ecology of Organizations." *American Journal of Sociology* 82 (March): 929–64.

Harris, M. 1968. *The Rise of Anthropological Theory*. New York: Crowell.

Hawley, A. 1950. *Human Ecology*. New York: Ronald Press.

_____. 1986. *Human Ecology: A Theoretical Essay*. Chicago: University of Chicago Press.

Hendershot, Gerry E., and Thomas F. James. 1972. "Size and Growth as Determinants of Administrative-Production Ratios in Organizations." *American Sociological Review* 37 (April): 149–53.

_____. 1961. *Social Behavior: Its Elementary Forms*. New York: Harcourt Brace Jovanovich.

_____. 1974. *Social Behavior: Its Elementary Forms*, rev. ed. New York: Harcourt Brace Jovanovich.

Hook, S. 1974. *From Hegel to Marx*. Ann Arbor: University of Michigan Press.

Hunter, Floyd. 1953. *Community Power Structure*. Chapel Hill: University of North Carolina Press.

Huntington, Samuel, and Jorgé Dominguez. 1975. "Political Development." In *Handbook of Political Science*, vol. 3, edited by Fred Greenstein and Nelson Polsby, 1–114. Reading, Mass.: Addison-Wesley.

James, Thomas F., and Stephen C. Finner. 1975. "System Size and Structural Differentiation in Formal Organizations." *Sociological Quarterly* 16 (Winter): 124–30.

Jenkins, J. Craig. 1983. "Resource Mobilization Theory and the Study of Social Movements." *Annual Review of Sociology* 9:527–53.

Johnston, William M. 1983. *The Austrian Mind: An Intellectual and Social History 1848–1938*. Berkeley: University of California Press.

Jones, Robert A. 1974. "Durkheim's Response to Spencer." *Sociological Quarterly* 15 (Summer): 341–58.

_____. 1986. *Émile Durkheim*. Newbury Park, Calif.: Sage.

Keller, Albert G. 1915. *Societal Evolution: A Study of the Evolutionary Basis of the Science of Society*. New York: Macmillan.

Kelley, Jonathan, and Herbert S. Klein. 1977. "Revolution and the Rebirth of Inequality." *American Journal of Sociology* 83:78–99.

Klapp, Orrin E. 1975. "Opening and Closing of Open Systems." *Behavioral Science* 20:251–57.

Kriesberg, Louis. 1982. *Social Conflict*. 2d ed. Englewood Cliffs, N.J.: Prentice-Hall.

Kuhn, Manford H., and T. S. McPartland. 1954. "An Empirical Investigation of Self-attitude." *American Sociological Review* 19 (February): 68–76.

Lenski, Gerhard. 1966. *Power and Privilege: A Theory of Stratification*. New York: McGraw-Hill.

———. 1970. *Human Societies.* New York: McGraw-Hill.

Lenski, Gerhard, and Jean Lenski. 1979. *Human Societies.* New York: McGraw-Hill.

Lukes, Steven. 1973a. *Émile Durkheim, His Life and Work.* London: Allen Lane.

———. 1973b. *Émile Durkheim, His Life and Work: A Historical and Critical Study.* New York: Penguin.

Mach, Ernst. 1938. *The Science of Mechanics.* Translated by T. J. McCormack. LaSalle, Ill.: Open Court.

Mack. R. W., and R. C. Snyder. 1957. "The Analysis of Social Conflict—Toward an Overview and Synthesis." *Journal of Conflict Resolution* 1 (June): 212–48.

Malinowski, Bronislaw. 1944. *A Scientific Theory of Culture and Other Essays.* Chapel Hill: University of North Carolina Press.

Marx, Karl [1857–1858] 1973. *Grundrisse.* New York: Vintage Books.

———. [1867] 1967. *Capital: A Critical Analysis of Capitalist Production.* Vol. 1, edited by F. Engels. New York: International.

Marx, Karl, and Friederich Engels. [1848] 1955. *The Communist Manifesto,* edited by S. Beer. New York: Appleton-Century-Crofts.

Mayhew, Bruce, and Roger Levinger. 1976. "Size and the Density of Interaction in Human Aggregates." *American Journal of Sociology* 82 (July): 86–110.

Mazuri, Ali, and Michael Tidy. 1984. *Nationalism and New States in Africa.* Nairobi, Kenya: Heinemann.

McCarthy, John D., and Mayer N. Zald. 1977. "Resource Mobilization and Social Movements: A Partial Theory." *American Journal of Sociology* 86:1212–41.

McHenry, Dean. 1979. *Tanzania's Ujamaa Villages: The Implementation of a Rural Development Strategy.* Berkeley, Calif.: Institute of International Studies.

Mead, George Herbert. 1934. *Mind, Self and Society.* Chicago: University of Chicago Press.

———. 1938. *The Philosophy of the Act.* Chicago: University of Chicago Press.

Merton, Robert K. 1957. *Social Theory and Social Structure.* New York: Free Press.

———. 1968. *Social Theory and Social Structure.* New York: Free Press.

Meyer, Marshall M. 1972. "Size and Structure of Organizations: A Causal Analysis." *American Sociological Review* 37 (August): 434–41.

Michels, Robert. [1915] 1959. *Political Parties.* Translated by Eden and Cedar Paul. New York: Dover.

Migdal, Joel S. 1974. *Peasants, Politics and Revolution.* Princeton, N.J.: Princeton University Press.

Miller, J. G. 1978. *Living Systems.* New York: McGraw-Hill.

Mills, C. Wright. 1959. *The Power Elite.* New York: Oxford University Press.

Montesquieu, Charles. [1748] 1900. *The Spirit of Laws.* London: Colonial Press.

Moore, Barrington. 1966. *Social Origins of Dictatorship and Democracy: Lord and Peasant in the Making of the Modern World.* Boston: Beacon Press.

Murdock, George P. 1940. "The Cross-Cultural Survey." *American Sociological Review* 5 (March): 361–70.

———. 1965. *Culture and Society.* Pittsburgh, Pa.: University of Pittsburgh Press.

———. 1967. *Ethnographic Atlas*. Pittsburgh: University of Pittsburgh Press.

Nisbet, R. A. 1966. *The Sociological Tradition*. New York: Basic Books.

———. 1974. *The Sociology of Émile Durkheim*. New York: Oxford University Press.

Nolan, Patrick D. 1979. "Size and Administrative Intensity in Nations." *American Sociological Review* 44 (February): 110–25.

Ollman, B. 1976. *Alienation: Marx's Concept of Man in Capitalist Society*. New York: Oxford University Press.

Paige, Jeffrey M. 1975. *Agrarian Revolution: Social Movements and Export Agriculture in the Underdeveloped World*. New York: Free Press.

Pareto, Vilfredo. [1901] 1968. *The Rise and Fall of Elites: An Application of Theoretical Sociology*. Introduction by H. Zetterberg. Totowa, N.J.: Bedminster Press.

———. [1909] 1971. *Manual of Political Economy*. Translated by A. Schwier and edited by A. Schwier and A. Page. New York: A. M. Kelley.

———. [1916] 1935. *Treatise on General Sociology*, edited by A. Livingston and translated by A. Bongiorno and A. Livingston. New York: Harcourt & Brace.

———. [1921] 1984. *The Transformation of Democracy*, edited by C. Powers and translated by R. Girola. New Brunswick, N.J.: Transaction.

Parsons, Talcott. 1937. *The Structure of Social Action*. New York: McGraw-Hill.

———. 1951. *The Social System*. New York: Free Press.

———. 1966. *Societies: Evolutionary and Comparative Perspectives*. Englewood Cliffs, N.J.: Prentice-Hall.

———. 1971. *The System of Modern Societies*. Englewood Cliffs, N.J.: Prentice-Hall.

Peel, J. D. Y. 1972. *Introduction in Herbert Spencer on Social Evolution*. Chicago: University of Chicago Press.

Perrin, Robert G. 1975. "Durkheim's Misrepresentation of Spencer: A Reply to Jones's 'Durkheim's Response to Spencer.'" *Sociological Quarterly* 16 (Autumn): 544–50.

———. 1976. "Herbert Spencer's Four Theories of Social Evolution." *American Journal of Sociology* 81 (May): 1339–60.

Popper, Karl. 1959. *The Logic of Scientific Discovery*. London: Hutchinson.

———. 1969. *Conjectures and Refutations*. London: Kegan Paul.

Powers, Charles. 1985. "On Regulatory Authority: Insights from Émile Durkheim." *Journal of the History of the Behavioral Sciences* 21:124–30.

Putnam, Robert, Robert Leonardi, Raffaella Nanetti, and Franco Pavoncello. 1983. "Explaining Institutional Success: The Case of Italian Regional Government." *American Political Science Review* 77:55–74.

Radcliffe-Brown, A. R. 1948. *A Natural Science of Society*. New York: Free Press.

Reynolds, Paul Davidson. 1971. *A Primer in Theory Construction*. Indianapolis: Bobbs-Merrill.

Ritzer, George. 1975. *Sociology and Multiple Paradigm Science*. Boston: Allyn & Bacon.

———. 1987. "The Current State of Metatheory." *Sociological Perspectives: The Theory Section Newsletter* 10:1–6.

_____. 1988. "Sociological Metatheory: A Defense of a Subfield by a Delineation of its Parameters." *Sociological Theory* 6:187–200.

_____. 1991. *Metatheorizing in Sociology*. Lexington, Mass.: Lexington Books.

Rose, Arnold M. 1967. *The Power Structure*. New York: Oxford University Press.

Rueschemeyer, Dietrich. 1977. "Structural Differentiation, Efficiency, and Power." *American Journal of Sociology* 83 (July): 1–25.

Schelling, Thomas C. 1960. *The Strategy of Conflict*. Cambridge, Mass.: Harvard University Press.

Schutz, Alfred. 1971. *Collected Papers*, vol. 3. The Hague: Nijhoff.

Schwartz, Barry. 1982. "The Social Context of Commemoration: A Study in Collective Memory." *Social Forces* 61:374–402.

Shelling, Thomas C. 1960. *The Strategy of Conflict*. Cambridge, Mass.: Harvard University Press.

_____. 1971. *The Strategy of Conflict*. New York: Oxford University Press.

Shibutani, Tamotsu. 1955. "Reference Groups as Perspectives." *American Journal of Sociology* 60 (May): 562–69.

_____. 1968. "A Cybernetic Approach to Motivation." In *Modern Systems Research for the Behavioral Scientist. A Source Book*, edited by W. Buckley. Chicago: Aldine.

Shils, E., and H. Finch, eds. 1949. *Max Weber on the Methodology of the Social Sciences*. New York: Free Press.

Simmel, Georg. 1890. *Ueber Soziale Differenzierung*. Leipzig: Duncker & Humblot.

_____. 1903–1904. "The Sociology of Conflict." Translated by A. Small. *American Journal of Sociology* 9:490–525, 672–89, 798–811.

_____. [1907] 1978. *The Philosophy of Money*. Translated by Tom Bottomore and David Frisby. Boston: Routledge & Kegan Paul.

_____. 1908. *Sociology: Studies in the Forms of Sociation*. Leipzig: Duncker & Humblot.

_____. [1908] 1956. "Conflict." In *Conflict and the Web of Group Affiliations*, translated by Kurt H. Wolff. New York: Free Press.

_____. 1950. *The Sociology of Georg Simmel: Earlier Essays*. Translated by Kurt H. Wolff. Glencoe, Ill.: Free Press.

Skocpol, Theda. 1979. *States and Social Revolution*. Cambridge: Cambridge University Press.

Smith, Adam. [1776] 1937. *An Inquiry into the Nature and Causes of the Wealth of Nations*. New York: Random House.

Sofranko, Andrew, and Robert Bealer. 1972. *Unbalanced Modernization and Domestic Instability: A Comparative Analysis*. Beverly Hills, Calif.: Sage.

Sorokin, P. A. 1961. "Variations on the Spencerian Theme of Militant and Industrial Types of Society." *Social Science* 36 (April): 91–99.

Spencer, Baldwin, and F. J. Gillen. 1899. *The Native Tribes of Central Australia*. London: Macmillan.

Spencer, Herbert. 1850. *Social Statics*. New York: D. Appleton.

_____. 1862. *First Principles*. New York: Appleton.

———. 1864–1867. *The Principles of Biology*. New York: Appleton.

———. 1873. *The Study of Sociology*. London: Kegan Paul.

———. 1873–1934. *Descriptive Sociology, or Groups of Sociological Facts*. (Published with Spencer's personal imprint and by various publishers.)

———. [1874–1896] 1898. *The Principles of Sociology*. New York: Appleton.

———. 1892–1898. *The Principles of Ethics*. New York: Appleton.

Spencer, Herbert, and David Duncan. 1874. *Types of Lowest Races*. London: Williams & Norgate.

Stephan, Edward G. 1971. "Variation in County Size: A Theory of Segmental Growth." *American Sociological Review* 36 (June): 451–61.

Stinchcombe, Arthur L. 1968. *Construction Social Theories*. New York: Harcourt.

Sumner, William Graham. 1906. *Folkways*. Boston: Ginn.

Sumner, William Graham, and Albert G. Keller. 1927. *The Science of Society*. New Haven, Conn.: Yale University Press.

Swanson, Guy E. 1961. "Mead and Freud: Their Relevance for Social Psychology." *Sociometry* 24 (December): 319–39.

Szymanski, Albert. 1973. "Military Spending and Economic Stagnation." *American Journal of Sociology* 79 (July): 1–14.

———. 1981. *The Logic of Imperialism*. New York: Praeger.

Thibaut, John W., and Harold H. Kelley. 1959. *The Social Psychology of Groups*. New York: Wiley.

Tilly, Charles. 1978. *From Mobilization to Revolution*. Reading, Mass.: Addison-Wesley.

Tilton, Timothy. 1975. *Nazism, Neo-Nazism, and the Peasantry*. Bloomington: Indiana University Press.

Turner, Jonathan H. 1972. *Patterns of Social Organization*. New York: McGraw-Hill.

———. 1973. "From Utopia to Where: A Critique of the Dahrendorf Conflict Model." *Social Forces* 52 (December): 236–44.

———. 1974. *The Structure of Sociological Theory*. Homewood, Ill.: Dorsey.

———. 1975. "A Strategy for Reformulating the Dialectical and Functional Theories of Conflict." *Social Forces* 53 (March): 433–44.

———. 1978. *The Structure of Sociological Theory*, 2d ed. Homewood, Ill.: Dorsey.

———. 1979a. "Sociology as a Theory Building Enterprise: Detours from the Early Masters." *Pacific Sociological Review* 72 (October): 427–56.

———. 1979b. "Toward a Social Physics." *Humboldt Journal of Social Relations* 7:123–39.

———. 1980. "Toward a Social Physics." *Humboldt Journal of Social Relations* 7 (Spring) 1: 140–55.

———. 1981. "Émile Durkheim's Theory of Interaction in Differential Social Systems." *Pacific Sociological Review* 24(4): 187–208.

———. 1984. *Social Stratification: A Theoretical Analysis*. New York: Columbia University Press.

———. 1985a. "In Defense of Positivism." *Sociological Theory* 3 (Fall): 24–30.

_____. 1985b. *Herbert Spencer: A Renewed Appreciation*. Newbury Park, Calif.: Sage.

_____. 1986. *The Structure of Sociological Theory*. 4th ed. Homewood, Ill.: Dorsey Press.

_____. 1987. "Analytical Theorizing." In *Social Theory Today*, edited by Anthony Giddens and Jonathan H. Turner, 156–94. Cambridge: Polity Press.

_____. 1988. *A Theory of Social Interaction*. Stanford, Calif.: Stanford University Press.

Turner, Jonathan H., and L. Beeghley. 1981. *The Emergence of Sociological Theory*. Homewood, Ill.: Dorsey Press.

Turner, Jonathan H., and Alexandra Maryanski. 1979. *Functionalism*. Menlo Park, Calif.: Benjamin-Cummings.

Turner, Ralph H. 1968. "Social Roles: Sociological Aspects." *International Encyclopedia of the Social Sciences*. New York: Macmillan and Free Press.

_____. 1978. "The Role and the Person." *American Journal of Sociology* 84 (July): 1–23.

_____. 1979. "A Strategy for Developing an Integrated Role Theory." *Humboldt Journal of Social Relations* 7 (1): 123–39 (Spring).

Turner, Stephen Park, and Jonathan H. Turner. 1990. *The Impossible Science: An Institutional Analysis of American Sociology*. Newbury Park, Calif.: Sage.

Useem, Michael. 1984. *The Inner Circle*. New York: Oxford University Press.

Wallace, Walter L. 1987. "Causal Images in Sociology." *Sociological Theory* 5 (Spring): 41–46.

Weber, Max. 1905. *The Protestant Ethic and the Spirit of Capitalism*. New York: Free Press.

_____. [1922] 1968. *Economy and Society: An Outline of Interpretive Sociology*, edited by G. Roth and C. Wittich. Berkeley: University of California Press.

Willer, David, and Murray M. Webster, Jr. 1970. "Theoretical Concepts and Observables." *American Sociological Review* 35: 748–57.

Williams, Robin M., Jr., 1947. *The Reduction of Intergroup Tensions*. New York: Social Science Research Council.

_____. 1970. "Social Order and Social Conflict." *Proceedings of the American Philosophical Society* 114 (June): 217–25.

Zetterberg, Hans. 1965. *On Theory and Verification in Sociology*. 3d ed. New York: Bedminster Press.

Index

"Abduction," 16
Adaptation, 31, 150–51, 152, 189
Adaptive capacity, 21
Aggregation, 19. *See also* Growth
Alienation, 9, 10, 92, 93, 145, 183
Anomie, 53, 60–61, 63–64, 65n, 68,
    70, 73, 77, 79, 81, 84n, 183,
    187–88
Anthropology, 35, 36, 42–43, 44, 45,
    47, 48, 54
"Associative and dissociative proc-
    esses," 88–90
Authority, 9, 10, 19, 110, 115–17,
    144–47
    centralization of, 20, 21, 22,
      52–53, 60–61
Axiomatic approach, 8, 12–13

Beeghley, Leonard, 147n
Behavioral structuralism, 153–54
Behaviorism, 156, 160
Blalock, Hubert M., 100n
Blumer, Herbert, 171
Bodin, 187
Boulding, Kenneth E., 100n
Bureaucracy(ies), 10, 186

Capitalism, 10, 18, 22, 101, 107
Causal model, 11–12, 13, 14, 48
Charisma, 112, 113
    charismatic leaders, 114, 145

The Charisma Principle, 124–25
Class, 9, 10, 84n, 110, 113, 116, 183,
    188
Collective conscience, 9, 52, 63, 76,
    79, 155
Collins, Randall, 6n, 55n, 78, 83,
    100n, 117
Communism, 101
The Communist Manifesto, 100n
Communities
    types of, 10
Competition, 51, 74, 137–38, 185
Complexity, 153
Comte, Auguste, 1, 5, 6n, 8, 12, 14,
    15, 24, 33n, 51, 68–69, 158n,
    159, 174, 177, 178–79, 180–82,
    184, 185, 189
Concept formation, 8, 9
Conflict, 73–74, 87–100, 101–7,
    109–17, 122–32, 135–47, 155,
    175, 178–79
    laws of, 87
*Conflict and the Web of Group Affil-
    iations*, 100n
The Conflict of Interest Principle,
    124–25
Cooley, Charles Horton, 152
Copresence, 74, 78, 79
Coser, Lewis A., 87, 100n, 132
Cours de philosophe positive, 1
Criminal gangs, 13, 175

Dahrendorf, Ralf, 87
Darwin, Charles, 10, 17, 65n, 75, 172
Das Kapital, 101
Decentralization, 22
Deduction, 16
Deinstitutionalization, 18
Delinquency, 10, 15
Dependence, 20, 186
The Deprivation Principle, 124–25
The Deprivation-Immiseration Principle, 124–25
Descriptive Sociology, 32–33n, 34n, 36, 37–38, 42, 43, 45, 45n, 181
Dewey, John, 150, 152
Differentiation, 16, 19, 20, 21, 22, 24, 25, 28, 31, 32, 62, 64, 64n, 68, 74, 76, 80–82, 136, 140, 154, 184
  cultural, 80
  evolutionary, 26
  internal, 29, 30
  societal, 23
  structural, 80–81
The Discontinuity-Tension Principle, 124–25
Disintegration, 51, 53–54, 77, 110, 184, 186
  sociocultural, 81
Dissolution, 19
Distribution, 91, 94, 126, 128, 145
  ecological, 32
  systems of, 63, 88, 91, 94
Diversification, 22
Division of labor, 31, 51, 53–54, 55n, 60–61, 69, 74, 75, 80, 84n
The Division of Labor, 49, 50, 52, 53, 54, 64n, 70, 74, 76, 78, 169, 181
Dominance relations, 15
Durkheim, Émile, ix, 2, 3, 8, 9, 12, 17, 24, 31, 32, 33n, 34n, 35, 39, 42, 44, 47–55, 55n, 57–65, 67–83, 84–85n, 126, 135–144, 146–147, 154–156, 158n, 168, 176, 177, 178–79, 181, 183, 184–88
"Durkheim's Laws," 67

Ecological concentration. *See* Human ecology
Economic development, 10, 13, 15, 175
Educational achievement, 15
Egoism, 53, 60–61, 64, 65n, 68, 70, 73, 77, 79, 82, 183, 187–88
Einstein, Albert, 48, 172, 183
Emotional arousal, 74, 78, 79, 93–96, 146
Empirical generalization, 8, 9–10, 11, 13, 14
Empiricism, 2, 4, 6n
Engels, Friedrich, 122, 136–41, 144–46, 186–87
Environment, 39, 40
  inorganic, 39, 40
  organic, 39, 40
  sociological, 39, 40
Ethnic groups, 10, 13, 175
Ethnic minorities. *See* Ethnic groups
Ethnographic analysis, 42
Evolution, 18, 20, 22, 33n, 36, 37, 50–51
  cosmic, 18, 20
  societal, 22–23, 24, 37
  theory of, 43
Evolutionism, 36
Exchange theory, 127, 130, 132, 174
The Expectation-Reciprocity Principle, 124–25
The Exploitation Principle, 124–25

The False-Consciousness Principle, 124–25
Family, 10, 13 175
  size, 10
Finance units in organizations, 13, 175
First Principles, 18, 31, 32n
Folkways, 36
French Revolution, 70
Freud, Sigmund, 43
Functionalism, 45
  explanatory, 44

Geopolitics, 114–17
Gestures
    "conversation of," 151, 162
    "significant," 151, 154, 155, 157,
        163, 164
Gibbs, Jack, 48, 172
Giddings, Franklin, 2–3
Gillian, F. J., 43
Goffman, Erving, 78
Growth, 21, 24, 25, 28, 32

Hegel, Friedrich, 2
Heterogeneity, 20
    coherent, 21
Historicism, 42, 45
History of ideas, 8–9, 47, 57–58, 100,
        132
Hobbes, Thomas, 87
Homans, George, 13, 175
Homogeneity, 20
Human ecology, 28, 51, 60–61, 62,
        70, 71–74, 80, 83n, 105, 137
Human Relations Area Files
        (HRAF), 35–45

Induction, 16
Industrialization, 10
Influence, 9
Institutionalization, 18, 19–20, 21
Integration, 16, 19, 20, 22, 28–29,
        47–55, 62–63, 68–71, 75, 89,
        109–17, 137–38, 153, 155,
        167–70, 184, 186–89
    sociocultural, 83
Interaction, 9, 16, 49–50, 74, 78,
        79–80, 82–83, 151, 160, 162,
        165–66, 172
    prolonged, in intimate contexts,
        15
Interactionism, symbolic, 149
Interdependence, 18, 82, 137

James, William, 152
Jones, Robert A., 57
The Justice Principle, 124–25

Keller, Albert G., 36, 43, 44

Labeling theory, 173–74
Laplace, Pierre Simon, 48, 172
Law of Conflict, 133
Law of Conflict Potential, 133
Law of Conflict Violence, 133
Lazarsfeld, Paul, 4
Legitimation, 111, 114
Lenski, Gerhard, 43

Mach, Ernst, 3
Mack, Raymond W., 100n
Maine, Henry, 24
Malinowski, Bronislaw, 42, 44
Malthus, Thomas, 87
Marital interaction, 15
Marx, Karl, ix, 2, 3, 8, 9, 12, 17, 24,
        32, 33n, 35, 48, 57, 58, 65n,
        87–100, 100n, 101–08, 110,
        122–33, 135–41, 144–47, 176,
        177, 178–79, 181–82, 183,
        184–88
Maxwell, 48, 172
Mead, George Herbert, ix, 8, 9, 12,
        32, 48, 149–58, 159–72, 176,
        177, 178–79, 182, 183, 184,
        188–89
Merton, 4, 13, 172, 175
Merton-Lazarsfeld. *See* Merton; La-
        zarsfeld
"Metaphysical stage" of sociological
        development, 5
Metasociology, 120
Metatheory, 67, 119–22, 123, 126,
        132–34, 171
Mill, John Stuart, 64n, 87
Mind, 152, 154, 156, 161–62,
        163–64, 165, 178–79
Mind, Self, and Society, 154
The Mobilization-Organization
        Principle, 124–25
The Mobilization-Solidarity Princi-
        ple, 124–25
Modeling procedure, 8, 11
Modernization, 10
Montesquieu, Charles Louis de, 53,
        71

Morgan, Henry Louis, 48
Movement of actors in hierarchies, 15
Murdock, George, 35–36

Natural laws, 1, 3, 69, 180
Natural science(s), 5, 7, 14, 47, 48, 120, 136, 175
  "of society," 136
Natural selection, 17, 50–51
Newton, Sir Isaac, 1, 48, 69, 71, 84n, 87, 172
Nonnormative behavior, 15

Occupational mobility, 12
On the Origin of Species, 65n
Operative processes, 39, 40

Pareto, Vilfredo, 37, 63, 135–44, 146–47, 177, 178–79, 181–82, 183, 185, 187–89
Park, Robert, 2
Parsonian action theory, 10, 174
Parsons, Talcott, 4, 17, 35, 58, 76, 175, 178–79
Parsons-Stouffer. *See* Parsons; Stouffer
Party, 110, 113, 116
Pathology(ies), 77
Pearson, Karl, 173
Perrin, Robert G., 57
The Philosophy of the Act, 161
Planck, Max, 48, 172
Political centralization, 10
Political democracy, 10
Political organization, 139–41, 147
Political oscillation, 141–44, 147
Political science, 136
Popper, Karl, 3
Positive philosophy. *See* positivism
Positivism, 1, 4, 5, 117, 119, 180–81
  logical, 4
Power, 9, 10, 88, 91, 105, 115, 135–47
  centralization of, 53, 142–44
  opposition to, 53, 105

The Power-Dependence Principle, 124–25
The Power-Use Principle, 124–25
Pragmatism, 150, 160
"Presentist" view, 62
Principles of Biology, 32n, 34n
Principles of Ethics, 32n
Principles of Psychology, 32n
Principles of Sociology, 22, 24, 30, 32n, 33n, 34n, 36, 37, 39, 58
Productivity, 10, 16, 51, 138, 184–86
  in human systems, 15, 137
The Protestant Ethic and the Spirit of Capitalism, 182

Radcliffe-Brown, A. R., 42, 44
The Rational-Choice Principle, 124–25, 127
The Rational-Manipulation Principle, 124–25
The Rebalancing Principle, 124–25, 127
Regulative processes, 39, 40
Relativism, 42, 45
Ritual, 74, 77, 78, 79, 80, 82–83
Ritzer, George, 120, 122, 132–33
Role-taking, 151–52, 154, 157, 163, 164, 166–70
Rousseau, Jean Jacques, 53
The Routinization Principle, 124–25

Saint-Simon, Comte de, 158n
Schelling, Thomas C., 100n
"Science of society," 36, 159
Scope, 153
Segregation, 19, 20, 21
Selection, 74
  mechanism of, 75
Self, 9, 152, 154, 156, 161–62, 164–65, 170–72, 178–79
Simmel, Georg, 17, 87–91, 95–100, 101, 102, 135–41, 144–47, 178–79, 181–82, 183
Skocpol, Theda, 117
Small, Albion, 2

Smith, Adam, 51, 58, 64n, 87, 122, 158n
Snyder, 100n
Social organization. *See* Social systems
"Social physics," 1, 6n, 7, 14, 15, 159–72, 174
Social processes, 13
Social psychology, 77
Social science(s), 37–38, 47, 48, 123, 176
Social Statics, 32n
Social systems, 10, 15, 18, 19, 20, 22, 25, 28–29, 33n, 43, 47–55, 57–65, 67–83, 87, 89, 96–99, 102–04, 110–12, 122, 137–39, 147, 166–68, 184–89
  industrial, 37, 62–63, 88
  militant, 62
  phases of, 22, 23
Socialization, 16, 172, 175
Societies
  compound, 24, 26–27
  simple, 24, 26–27
Sociological Theory, 84n
Sociology, defined, 1
  of conflict, 87–100
  political, 135, 143, 147
  quantitative, 3
  scientific, 7, 69, 101, 109, 132
Solidarity, 70, 78, 83, 97
  mechanical, 63
Specialization, 75
Spencer, Baldwin, 43
Spencer, Herbert, 2, 8, 9, 12, 17–32, 32–34n, 35–45, 48, 51, 57–65, 76, 135–44, 146–47, 177, 178–79, 180–82, 183, 184–87, 188–89

Status group, 110, 113, 116
Stinchcombe, Arthur L., 71
The Structure of Social Action, 35
The Study of Sociology, 32n, 33n
Suicide, 54, 65n, 70, 84n, 181
Sumner, William Graham, 2, 36, 43, 44
"Survival of the fittest," 65n
Symbolic interactionism, 149, 171, 174
Symbolic logic, 4
"Synthetic Philosophy," 33n

Technology, 51, 70, 73, 138
Tocqueville, Alexis de, 53, 71
Tonnies, Ferdinand, 24
Tyler, Edward E., 24

Utilitarianism, 123, 150, 188–89

Verstehen, 9
Vienna Circle, 2, 3, 5
Violence, 89, 90, 93–99, 106–07, 127, 131–33, 146
The Violence-Redistribution Principle, 124–25

Wallace, Walter, 84n
Ward, Frank Lester, 2
Watson, John B., 160–61
The Wealth of Nations, 158n
Weber, Max, ix, 3, 9, 17, 24, 32, 109–17, 122–33, 135–41, 144–47, 176, 178–79, 181
Williams, Robin M. Jr., 100n
The Withdrawal of Legitimacy Principle, 124–25
Wolff, Kurt H., 100n
Wundt, Wilhelm Max, 151